QUANTITATIVE GENETIC VARIATION

ACADEMIC PRESS RAPID MANUSCRIPT REPRODUCTION

QUANTITATIVE GENETIC VARIATION

Edited by

JAMES N. THOMPSON, JR.
Department of Zoology
University of Oklahoma
Norman, Oklahoma

J. M. THODAY
Department of Genetics
University of Cambridge
Cambridge, England

Academic Press   New York   San Francisco   London   1979
A Subsidiary of Harcourt Brace Jovanovich, Publishers

COPYRIGHT © 1979, BY ACADEMIC PRESS, INC.
ALL RIGHTS RESERVED.
NO PART OF THIS PUBLICATION MAY BE REPRODUCED OR
TRANSMITTED IN ANY FORM OR BY ANY MEANS, ELECTRONIC
OR MECHANICAL, INCLUDING PHOTOCOPY, RECORDING, OR ANY
INFORMATION STORAGE AND RETRIEVAL SYSTEM, WITHOUT
PERMISSION IN WRITING FROM THE PUBLISHER.

ACADEMIC PRESS, INC.
111 Fifth Avenue, New York, New York 10003

United Kingdom Edition published by
ACADEMIC PRESS, INC. (LONDON) LTD.
24/28 Oval Road, London NW1 7DX

Library of Congress Cataloging in Publication Data

Main entry under title:

Quantitative genetic variation.

    Includes index.
    1. Quantitative genetics.   2. Variation (Biology)
I. Thompson, James N.   II. Thoday, J. M.
[DNLM: 1. Biometry—Congresses.   2. Variation
(Genetics)—Congresses.   QH323.5 Q15]
QH452.7.Q36   575.2   79-9917
ISBN 0-12-688850-7

PRINTED IN THE UNITED STATES OF AMERICA
79 80 81 82   9 8 7 6 5 4 3 2 1

CONTENTS

Contributors	vii
Preface	ix
Acknowledgments	xi

PART I INTRODUCTION TO THE STUDY OF QUANTITATIVE GENETIC VARIATION

Introduction   1
James N. Thompson, Jr., and J. M. Thoday

Historical Overview: Quantitative Variation and Polygenic Systems   5
Kenneth Mather

Quantitative Genetic Variation in Fungi   35
C. E. Caten

Polygenic Variation in Natural Populations of *Drosophila*   61
P. A. Parsons

PART II THE BIOMETRICAL APPROACH TO QUANTITATIVE VARIATION

The Biometrical Approach to Quantitative Variation   81
J. L. Jinks

PART III THE USES AND LIMITATIONS OF SELECTION

An Overview of Selection Theory and Analysis   111
Clifford Johnson

Computer Simulations — 121
  H. Geldermann and H. Gundel

Canalization and Selection — 139
  J. M. Rendel

The Posterior Crossvein in *Drosophila* as a Model Phenotype — 157
  Roger Milkman

PART IV ANALYSIS OF INDIVIDUAL GENE EFFECTS

Polygenic Mutations — 177
  Terumi Mukai

Uses of Recombinant Inbred Lines — 197
  Alberto Oliverio

Polygene Mapping: Uses and Limitations — 219
  J. M. Thoday

Computer Simulation of the Breeding Program for Polygene Location — 235
  James N. Thompson, Jr., and Timothy N. Kaiser

Polygenic Influences Upon Development in a Model Character — 243
  James N. Thompson, Jr.

Genes Affecting Quantitative Aspects of Physiology in Rodents — 263
  John G. M. Shire

Cytological Markers and Quantitative Variation in Wheat — 275
  C. N. Law and M. D. Gale

Synthesis: Polygenic Variation in Perspective — 295
  James N. Thompson, Jr., and J. M. Thoday

INDEX — 303

Contributors

Caten, C. E. (35), Department of Genetics, University of Birmingham, Birmingham, United Kingdom

Gale, M. D. (275), Plant Breeding Institute, Cambridge, England

Geldermann, H. (121), Institut für Tierzucht und Vererbungsforschung, Tierärztliche Hochschule, Hannover, Germany

Gundel, H. (121), Institut für Tierzucht und Vererbungsforschung, Tierärztliche Hochschule, Hannover, Germany

Jinks, J. L. (81), Department of Genetics, University of Birmingham, Birmingham, United Kingdom

Johnson, Clifford (111), Department of Zoology, University of Florida, Gainesville, Florida

Kaiser, Timothy N. (235), Department of Zoology, University of Oklahoma, Norman, Oklahoma

Law, C. N. (275), Plant Breeding Institute, Cambridge, England

Mather, K. (5), Department of Genetics, University of Birmingham, Birmingham, United Kingdom

Milkman, Roger (157), Department of Zoology, The University of Iowa, Iowa City, Iowa

Mukai, Terumi (177), Department of Biology, Kyushu University, Fukuoka, Japan

Oliverio, Alberto (197), Istituto di Fisiologia Generale, University of Rome, and Laboratorio Psicobiologia e Psicofarmacologia, Rome, Italy

Parsons, P. A. (61), Department of Genetics and Human Variation, La Trobe University, Bundoora, Victoria, Australia

Rendel, J. M. (139), CSIRO Division of Animal Production, North Ryde, NSW, Australia

Shire, John G. M. (263), Institute of Genetics, University of Glasgow, Glasgow, Scotland

Thoday, J. M. (1, 219, 295), Department of Genetics, University of Cambridge, Cambridge, England

Thompson, James N., Jr. (1, 235, 243, 295), Department of Zoology, University of Oklahoma, Norman, Oklahoma

Preface

The study of quantitative genetic variation involves a fusion of ideas from many areas of genetics. Although some of these, such as the mathematical-statistical analysis of variation, have received the bulk of attention because of their value in applied biology and evolutionary theory, interest in quantitative variation in other contexts is growing. It is becoming clear, for example, that the stepwise phenotypic changes that characterize selection responses can provide an important insight into the rate and range of developmental responses, and thus provides a powerful, though seldom tapped, resource for studies in developmental genetics. The information available from quantitative genetics complements not only our understanding of evolutionary processes, but also our interpretation of the genotype–phenotype relationships basic to all genetic questions.

Several recent conferences and symposia, including a symposium on the nature of polygenes held at the Department of Genetics of the University of Cambridge, convinced us that it would be valuable to bring together a number of papers describing and discussing some of the major experimental approaches to quantitative genetic variation. Although this is not the proceedings of any particular symposium, its makeup was influenced by our interactions with many individuals at these meetings.

The primary objective of the book, therefore, is to present a survey of some of the experimental approaches to quantitative genetic variation, summarize the major contributions of each approach, and discuss their potential applications and limitations.

The first section provides a historical overview of the field and descriptions of the assessment of genetic variation in two groups, fungi and *Drosophila*. The contrast between these two organisms illustrates the point that in both the commonly used genetic organism *Drosophila* and (at least from the quantitative genetic point of view) the less often studied fungi, there are tremendous opportunities for analyzing the polygenic structure of natural populations and its physiological and ecological significance. A discussion of one of the most widely applicable tools, biometrical analysis, is designed for a nonspecialist audience. This is followed by a brief survey of artificial selection, showing the insights that selection experiments can give to specific developmental and genetic questions.

Almost half of the book concentrates on the analysis of individual polygenic effects. A wide variety of techniques now allow one to focus on a single chromosome, chromosome region, or polygenic locus. The scope of the chapters in the last section intentionally varies from a broad to a fairly restricted assessment of a technique or its application. The whole section gives a guide to the literature and a good overview of our current ability to identify and manipulate specific components of a quantitative genetic trait.

Quantitative genetics is a dynamic subject with a long and complex history. It is our hope that the methods and conceptual approaches discussed here will encourage a fresh look at some of the problems and untapped potentials of the study of polygenic variation.

Acknowledgments

In planning this book, we are grateful for the encouragement given by a number of individuals, foremost among them being Professor Sir Kenneth Mather. We also appreciate the care and patience shown by Ms. Marian Cothran and Mr. Joe V. Toney in proofreading and indexing.

*Introduction to the Study
of Quantitative Genetic Variation*

INTRODUCTION

James N. Thompson, jr.

Department of Zoology
University of Oklahoma
Norman, Oklahoma

J. M. Thoday

Department of Genetics
University of Cambridge
Cambridge, England

Quantitative genetics is, at the same time, one of the best and one of the poorest understood areas of genetics. The techniques of artificial selection have been used by plant and animal breeders for centuries. Elegant biometrical models of additive, dominance, and interaction components of variation, of selection responses, and of the genetical structure of natural populations have been tested and refined. Yet, our understanding of the nature and function of the specific genes determining the genetic component of variation in quantitative traits is still very poorly understood.

The history of quantitative genetics is to a large extent separated from the history of traditional Mendelian genetics by methodology. The ability to classify phenotypes of discontinuous traits into discrete groups and make counts of the number of individuals in each group leads to statistical analyses employing the χ^2 test of goodness of fit. The continuous distributions that characterize quantitative traits, on the other hand, lead to analyses that compare the variances between and within groups and, thus, to a biometrical approach.

The mathematical foundations of a study of quantitative variation, first initiated by Galton's application of mathematics to biological problems (1889), had been begun before the rediscovery of Mendel's work. It remained to others, notably Johannsen (1909), Nilsson-Ehle (1909), East (1910), and Fisher (*e.g.*, 1918), to uncover the specific relationships between Mendelian and early biometrical approaches that formed the primary basis for an understanding of quantitative genetic variation.

In particular, Johannsen (1909), in demonstrating that both heritable and non-heritable factors were responsible for the variations observed in the seed weight of beans and that contributions of these factors could be distinguished in breeding tests, provided a major insight into the relationship between the genotype and the phenotype. Then, Nilsson-Ehle (1909), investigating grain color in wheat, discovered that hereditary factors existed whose actions in the determination of the phenotype were similar, if not identical, and realized that a number of such factors, if each had smaller individual actions, could account for the phenotype observed in the continuous variation of quantitative characters.

Mather (1941) spoke of "polygenes" in the context of biometrical genetics as a term to describe the genes contributing to continuous or quantitative phenotypic variation. As a simplifying assumption of biometrical models, polygenes were said to have similar phenotypic effects that were small, relative to other sources of variation. Large numbers of polygenes were assumed to affect each trait.

Unfortunately, some have come to think of these as established facts, applicable to all examples, rather than as simplifying assumptions. One consequence of this misunderstanding is the common belief that polygenes are always too small in effect and too numerous to be handled individually. Without individual manipulation, attempts to analyze developmental influences of polygenic loci are not possible. Mather certainly never intended such limitations to be ascribed to the studies of polygenic systems, and the description of an effective technique for isolating polygenes by Thoday (1961) and similar approaches to dissecting quantitative traits (*e.g.*, Law, 1966; Milkman, 1970) opened polygenic characters to precise developmental analysis (Spickett, 1963). Such investigations have shown that the factors responsible for heritable differences of a continuous variable may sometimes be few, unequal in effect, and different in action (see Thompson and Thoday, 1974, and Thompson, 1975, for reviews). If this book is biased, it is because it gives unusual stress to the techniques appropriate in such situations, though we have not neglected to give room for discussion of the more generally applicable techniques of Biometrical Genetics.

Controversy concerning the number of genes involved in continuous phenotypic distributions, and hence the applicability of location techniques, sometimes arises because of confusion between two of the questions that one can ask about a continuous variable but that have quite different answers. The first question concerns the number of loci (which may sometimes be small) segregating in the particular material under study. The other question concerns the number of genes in a species' genome or gene pool at which variation can affect the variable; the number of such loci is probably always large. The same two questions can, of course, be asked about discontinuous variation: for instance over 70 loci are known in *Drosophila melanogaster* at which segregation gives discontinuous eye color phenotypes, but seldom do more than one or two (except in specially bred stocks) segregate in the same material. That 70 loci can contribute to the formation of structure and pigment does not make eye color a polygenic character. The critical question concerns the number of *segregating* loci in a particular experiment or population, always recognizing that continuous variation arises not only from number of loci but also from limitations to the *relative* effects of specific allelic substitutions.

One should always beware against confusing these questions (see Thoday and Thompson, 1976). Answers to the first question, particularly when they involve linkage map locations, can contribute to answers to the second. But answers to the second that invoke or prove that many loci may be effective should not prevent us from asking whether any loci in our specific research material can be handled individually.

Much of this book is concerned with techniques for the study of specific limited segregations. This is because we believe that these may lead us to more precise information about the nature (or variety of natures) of polygenes, the way they respond to selection, how polygenic systems are organized in specific populations, how they act in development, and how analysis of continuous variation can throw light on developmental problems.

We hope that the diversity of papers included will help to stimulate advances in our understanding of polygenic systems, their component genes, and the roles they play in development and evolution.

REFERENCES

East, E.M. (1910). *Amer. Nat.* 44, 65-82.
Fisher, R.A. (1918). *Trans. Roy. Soc. Edinburgh* 52, 399-433.
Galton, F. (1889). "Natural Inheritance." Macmillan, London.
Johannsen, W. (1909). "Elemente der exakten Erblichkeitslehre." Fischer, Jena.

Law, C.N. (1966). *Genetics 53*, 487–498.
Mather, K. (1941). *J. Genet. 41*, 159–193.
Milkman, R.D. (1970). *Adv. Genet. 15*, 55–114.
Nilsson-Ehle, H. (1909). "Kreuzunguntersuchungen an Hafer und Weizen." Lund.
Spickett, S.G. (1963). *Nature 199*, 870–873.
Thoday, J.M. (1961). *Nature 191*, 368–370.
Thoday, J.M., and Thompson, J.N., Jr. (1976). *Genetica 46*, 335–344.
Thompson, J.N., Jr. (1975). *Nature 258*, 665–668.
Thompson, J.N., Jr., and Thoday, J.M. (1974). *Heredity 33*, 430–437.

HISTORICAL OVERVIEW:
QUANTITATIVE VARIATION AND POLYGENIC SYSTEMS

Kenneth Mather

Department of Genetics
University of Birmingham
Birmingham, U.K.

I. THE GENETICAL INTERPRETATION

Between Mendel's presentation of his experiments with peas to the Brünn Natural History Society in 1865 and the recognition in 1900 of the basic importance of his findings, Francis Galton had also attempted to elucidate the principles of hereditary transmission and had published the results in his book Natural Inheritance (1889). Galton's approach differed from Mendel's in two fundamentals. First, influenced no doubt by the emphasis his cousin Charles Darwin laid on the importance of small cumulative steps in evolutionary change, he did not follow the transmission of sharply contrasting characters as Mendel did, but chose to observe metrical or quantitative variation, in which individuals show every gradation of expression of the character between wide extremes, and which even to the most casual observer is rife in living species especially in the most familiar of all -- man himself. Secondly, he did not start off with crosses between true-breeding lines and follow the disappearance and reappearance of a character in the derived generations, but sought by the use of his correlations and regressions, to measure the average similarity between pairs of individuals of known relationship, parent and offspring or sib and sib, over the wide spectrum of the population itself. One can appreciate the consideration that must have led him to regard this as a profitable approach to adopt; but as we can now see, it was doomed to failure from the start. It would not bring out clearly the distinction between determinant and effect, between the genotype and the phenotype; and although Galton had the idea that the hereditary materials might be particulate, his observations could not enable him to recognise the individual hereditary factors, or to distinguish clearly the basic phenomena of segregation and recombination. He was able, however, to show that the correlation measuring hereditary similarity, was the same between father and offspring as between mother and offspring, so leading to the conclusion that hereditary transmission must be equilinear from the two parents -- a conclusion that in an elaborated form, was enshrined in his so-called Law of Ancestral Heredity. At the same time, despite his lack of success in establishing the principles of hereditary transmission, we owe a great debt to Galton, for he had shown us how quantitative variation and co-variation could be measured and handled, and in doing so had given us the basis for fruitful analysis in the future.

Unlike Mendel's work, Galton's did not pass unnoticed. By 1900 it was being expanded vigorously by Karl Pearson and his associates, with the result that when Mendel's Principles came to the fore there was a ready-made opposition to the idea that they were of general significance. This opposition was aggravated, too, by an underlying divergence of view on the relative importance of Galton's smooth variation and Mendel's sharply discontinuous differences for the understanding of adaptation and evolutionary change. The ensuing controversy between the Mendelians led by Bateson, and the Biometricians led by Pearson and Weldon, had its entertaining aspects, as R.C. Punnett (1950), himself one of the Mendelian protagonists, describes. But it was bitterly polemical and led nowhere, least of all to any understanding of the genetical component in quantitative variation. Resolution of the problem came in a different way.

In 1909, Johannsen published a detailed account of his observations on seed weight in the bean *Phaseolus vulgaris*, which resembles Mendel's peas in that the flowers naturally self-pollinate so giving rise to true-breeding or pure lines. Johannsen's study showed that seed weight displayed quantitative variation, that this variation could spring from non-heritable causes as well as heritable factors, that variation from the two causes could not be distinguished by inspection but only by breeding tests, that the two kinds of difference could both reinforce and oppose one another, and that there must be more than one heritable factor involved. At the same time, Nilsson-Ehle (1909) showed in cereals that differences large enough to be followed by Mendel's technique, and so traceable to single Mendelian factors, could be additive in their effects on the character of the individual. The basis was, thus, laid for the multifactorial (or polygenic as we may now say) interpretation of the genetical component of quantitative, metrical or continuous variation (as it has been variously styled), which was put forward by Nilsson-Ehle and independently by East. In essence this ascribed quantitative variation to the simultaneous action of a number of Mendel's factors (or genes as they soon came to be called) and of non-heritable agencies, the effects of the gene differences being too small when seen against the background of the overall variation for their segregations to be identified individually by the Mendelian technique, yet capable of adding to or supplementing one another. The expression of the character thus depends on the balance between the numbers of genes pulling in the two directions, towards enhancement and diminution, of the expression rather than on the specific effects of the genes themselves, with the non-heritable variation serving to blur any residual discontinuities in the variation to which the gene system gave rise. In other words quantitative variation such as Galton observed sprang from systems of genes, each inherited in the Mendelian fashion, but having effects on the expression of the character similar to one another. These effects were so small in relation to the total variation, heritable and non-heritable, as to be incapable of individual recognition by the Mendelian technique, while at the same time being capable of supplementing one another and so combining to produce large differences of phenotype.

II. ESTABLISHING THE INTERPRETATION

To put forward such an interpretation is one thing. To demonstrate its validity is another. By the genetical standards of the time it was a complicated interpretation and it was treated even by geneticists themselves sometimes with outright rejection and frequently with a reservation that amounted to suspicion -- a suspicion that lingered on at least for thirty years (see for example 'Espinasse, 1942). Closer analysis and more direct evidence was required, and this began to appear in the second decade of the century. It came in two ways. Galton and Pearson calculated their correlations between relatives from observations made not on experimentally bred families but on naturally breeding populations. In a paper whose sweep, mastery and innovation have deservedly made it a classic, Fisher (1918) showed that the biometrical results *must* follow if the genetical determinants were inherited in the Mendelian fashion. He demonstrated the occurrence of Mendel's phenomenon of dominance in quantitative variation; he showed how to allow for assortative mating; and he considered the consequences of linkage and certain forms of gene interaction. He put the genetical analysis of quantitative variation on an entirely new, more precise and more informative footing; and incidentally in partitioning the variation numerically into terms dependent on the different causal agents, he laid the foundation for the analysis of variance (now often abbreviated, somewhat inelegantly, to ANOVA) which has proved so powerful a tool in the statistical reduction of many kinds of data.

The second way in which new evidence was obtained was statistically less sophisticated but genetically was akin to Mendel's own approach. All the variation within a true-breeding line must be non-heritable. The reciprocal F_1's from crosses between two such lines should be alike and all their variation should also be non-heritable. But the genes are expected to segregate in F_2 and so add a heritable component to the non-heritable. The variation in F_2 should, therefore, be greater than in the parental and F_1 generations. Furthermore in the F_2 half the genes will be homozygous and half heterozygous, though the individuals will differ from one another in the numbers of their genes falling into the two classes. So while the average genetic variation of the F_3 families should be half that of the F_2, the variation will not be the same in all the F_3's. Furthermore, because of recombination the F_2 individuals must differ in the genes for which they are homozygous and the average expression of the character in each F_3 family will, therefore, be correlated with the expression of the F_2 individual from which it was derived. Thus, by generalising the effects of segregation and recombination we can arrive at expectations for the behaviour of variation in a Mendelian cross. A number of early experiments of this general kind were carried out by East and his collaborators (see for example East, 1915; Emerson and East, 1913). The results were in satisfactory agreement with the expectations, so confirming that the principles of segregation and recombination appied to quantitative variation as well as to the sharply contrasting differences from whose study Mendel originally derived them. Many such experiments have been carried out in later years, all of which have borne out, and indeed elaborated, the early findings (see Mather and Jinks, 1971).

III. CHROMOSOMES AND QUANTITATIVE VARIATION

As a result of the *Drosophila* investigations, by 1920 the chromosome theory had become generally, if by then not universally, accepted. The phenomenon of segregation was seen as arising from the behaviour of the chromosomes on which the genes (as Mendel's factors had come to be known) were borne, and the phenomena of linkage and crossing-over had been recognised. This opened up new possibilities for the study of quantitative variation: to the extent that its genetical component depended on genes carried on the chromosomes in the same way as the genes traceable by the Mendelian technique, its behaviour should reflect these same phenomena albeit in its own way. Furthermore in *Drosophila* easily recognisable genes were already available which could be used not only to mark all of the four chromosomes, but to mark specific pieces of the chromosomes and so enable them to be followed in experiment. These marker genes could, therefore, be used to investigate the contributions that particular chromosomes and pieces of chromosomes made to the determination of quantitative variation.

Interestingly enough the first evidence of linkage between the major gene differences, traceable by the Mendelian method, and the determinants of quantitative variation came not from *Drosophila* but from beans. Sax (1923) reported a cross between two strains of *Phaseolus vulgaris*, one of which had large coloured seeds and the other small white ones. The colour difference proved to depend on a single gene difference behaving just like the difference in testa colour that Mendel investigated in his peas. Seed size on the other hand showed continuous variation in accordance with Johannsen's earlier findings. In F_2 the plants homozygous for the allele giving coloured seed had an average seed weight greater than that of the plants homozygous for the white allele, with the average weight of the seed from beans heterozygous for the coloured alleles being just about midway between the two. Thus bean size was nearly proportional to the number of coloured alleles present in the F_2 individual, as would be expected if one or more genes affecting seed size were linked to the colour locus. This is not, of course, final evidence of linkage, since it could be due to the colour gene having a pleiotropic action on seed size. A similar kind of experiment in peas, made later by Rasmusson (1935) using coloured *v.* white flowers as the marker gene and flowering time as the quantitative character, put the issue beyond doubt. Rasmusson was able to extract from his progenies individuals in which the parental association of colour and flowering time had been reversed, and reversed by recombination, as later breeding tests clearly showed.

The wealth of marker genes and the special stocks which rapidly became available in *Drosophila*, however, made possible more comprehensive and more informative experiments than could be undertaken in other species. It was early shown by the use

of appropriate markers that all the four chromosomes carried
genes affecting egg size (Warren, 1924). Furthermore the
possibility of a pleiotropic effect could be ruled out by the
varying differences revealed when the effects on the quantitative character of a number of unmarked chromosomes were compared with the same marked homologue. A technique was developed by which it was possible simultaneously to assay the
contributions that at least the three major chromosomes (X, II
and III) were making to the quantitative differences in a
character between two or more stocks (Mather, 1942). Its use
showed that the sum of the effects traceable to the chromosomes effectively accounted for the whole of the differences
among lines of *Drosophila* in respect of the quantitative variation displayed by the number of chaetae borne on certain of
the abdominal segments (Mather and Harrison, 1949). The technique of chromosome assay has not only been refined and elaborated over the years in *Drosophila* (see Thoday, 1961; Cooke
and Mather, 1962), but more recently it has been extended to
wheat where methods of a different kind are available for the
manipulation of identifiable chromosomes (Law, 1967). Whenever adequately comprehensive assays have been made the result
has been the same: not only do nuclear genes contribute to
continuous or quantitative variation, they account for its
genetical component as nearly completely as they do for the
discontinuous or qualitative variation that can be analysed
directly by the classical methods of genetics. Galton's
variation traces to the chromosomes in just the same way and
to just the same extent as does Mendel's. This finding has a
number of consequences and raises a number of questions to
which we must return later. First, however, we must look at
the genetical analysis of quantitative variation to which it
leads in species where special facilities of the kind we have
been discussing are not available.

IV. BIOMETRICAL GENETICS

The use of the powerful Mendelian technique of genetical
analysis depends on the ability to assign the members of a
family to clearly delimited phenotypic classes which can be
interpreted unambiguously in terms of the genotypes relevant
to the analysis in hand. This is, of course, impossible with
the quantitative variation we are discussing, because the
segregation of individual gene-pairs is obscured by the variation arising from both the segregation of all the other genepairs affecting the character in question and the effects of
non-heritable agencies. Two courses are then open. One is to

seek to lessen the obscurity by reducing the number of gene-pairs segregating and reducing the variation stemming from non-heritable agencies. This latter may be attempted by standardising the environmental conditions; but complete success cannot be expected from doing so because much of the non-heritable variation stems not from external factors but from instability in the development of the organism (Mather, 1953b). Progeny testing offers a more effective way, however, because the mean expression of the characters in a family will be subject to much less non-heritable variation than that of an individual, no matter what the source of this variation may be. Reduction in the number of segregating gene-pairs may be attempted by inbreeding, but there must always remain doubt as to which of the gene-pairs are continuing to segregate in a given inbred line unless there are means available for distinguishing them. Such means are afforded in *Drosophila* by the linkage relations with marker genes, and a very effective method of isolating and specifying genes contributing to quantitative variation by means of their linkage relations was developed by Thoday (1961) using a combination of marker genes and progeny testing. Valuable as it is, however, this approach makes demands for marker genes, for progeny testing, for generations and therefore time, and for numbers which can seldom if ever be met among higher organisms outside *Drosophila*.

The second course is not restricted in its use by these requirements. This approach was pioneered by Fisher (1918) in his classical paper to which reference has already been made. It consists essentially of taking biometrical quantities, such as Galton used to measure the degrees of similarity among relatives, and analysing them in Mendelian terms -- by in fact combining Galton's biometrics with Mendel's genetics. Thus, using not Fisher's original notation but a later one which he also initiated (Fisher $et\ al.$, 1932), if we denote the increments added to the character by the genotypes AA, Aa, and aa by "d", "h", and "-d" (all measurements being taken from the mid-point between the two homozygotes) Fisher showed that in a randomly breeding population this gene-pair will contribute $2uv[d + (v-u)h]^2$ to the additive variation of the population and $4u^2v^2h^2$ to the variation traceable wholly to dominance, where u and v (= 1 - u) are the gene frequencies of A and a, respectively. Now given that the different gene-pairs are independent of one another in both their distribution in the population and in the increments they add to the character, their contributions to the additive and dominance components of variation will be additive and we can express the two components as $2\Sigma\{uv[d + (v-u)h]^2\}$ and $4\Sigma(u^2v^2h^2)$, respectively, summation proceeding over all genes affecting the variation of the character. There will, of course, also be a component of variation arising from non-heritable effects. If we now write

$$D_R = 4\Sigma\{uv\,[d + (v-u)h]^2\} \quad \text{and} \quad H_R = 16\Sigma(u^2v^2h^2)$$

and use E to represent the non-heritable variation, the total variation of the character, as measured by the variance, is

$$V = \tfrac{1}{2}D_R + \tfrac{1}{4}H_R + E.$$

We can similarly obtain expressions in the same terms for the co-variances between for example parent and offspring in the form

$$W_{po} = \tfrac{1}{4}D_R,$$

and full-sibs, in the form

$$W_{ss} = \tfrac{1}{4}D_R + 1/16\,H_R.$$

Galton's correlations between parent and offspring will, thus, be

$$r_{po} = \tfrac{1}{4}D_R / (\tfrac{1}{2}D_R + \tfrac{1}{4}H_R + E)$$

and that between sibs

$$r_{ss} = (\tfrac{1}{4}D_R + 1/16\,H_R)/(\tfrac{1}{2}D_R + \tfrac{1}{4}H_R + E).$$

We are then clearly on the way to estimating the genetical quantities D_R and H_R from the biometrical statistics, V, W, and r.

When the analysis is applied to data from human populations, such as Galton and Pearson observed, complications enter in. Some of these such as assortative mating were considered by Fisher; but others, such as the effects of the family environment, were not. It is not, however, our concern to discuss further the analysis and its complications, but to see the essential nature of the approach and to note certain of its characteristic features. The concepts and methodology of biometrical genetics will be taken further by Professor Jinks in a later section of this book

In the first place the biometric analysis is not selective as the Mendelian is: it does not concentrate on particular genic aspects of the variation but covers the whole of it, the relevant effects of every gene-pair that affects the expression of the character being included in the components of variation. The biometrical analysis, thus, offers the advantage of measuring and analysing the totality of the variation, which Mendelian analysis does not. But it equally carries the limitation of providing no means of distinguishing the effects of the individual genes one from another, which Mendelian analysis does. Thus biometrical and Mendelian analyses have complementary properties: the one covers the whole sweep of variation without isolating, or needing to isolate, the individual effects of the genes from which it stems; whereas the other isolates the individual contributions where these are large enough to be isolatable, but inevitably leaves out of account genes of lesser effect. The Mendelian analyses will reveal precise detail in the parts of the picture to which it can be applied, and for this reason will generally be preferred wherever its use is made possible by the nature of the variation, the material and the facilities available. The biometrical analysis will delineate the whole picture, albeit in less specific detail, and will be used wherever there is need to take all genic sources into account, or where either the nature of the variation iself, or considerations of experimental economy puts Mendelian analysis out of court. Thus the two are not rival ways of doing the same job: each has its own place and purpose. We might note, too, that the biometrical analysis of continuous variation, for which it was developed and for which it is specifically suited, carries no implication about the number of genes segregating: it will detect the lack of a genetical component in the variation of an inbred line as well as measure that component no matter how many genes are contributing to it. Nor does the analysis require that the frequency distribution representing the variation follows the Normal curve, as Galton assumed it to do. The analysis makes no assumptions about this and indeed it can be used even where the distribution shows discontinuities arising from genic differences, though in such a case the Mendelian analysis would generally be preferred as more economical of data and more specific in its results.

Galton and Pearson's observations which Fisher analysed were taken from an open-bred population. Such data, however, have limitations. One will already be obvious: though the additive component of variation provides the basis for predicting advances under selection (see Falconer, 1960) it depends not just on "d", measuring the differences between allelic homozygotes but also on "h", reflecting dominance. The effects of the allelic genes in homozygous and heterozygous states have not in fact been separately assessed. Nor can they be from such data, and the reason is that in a population we cannot "arrange the forms with certainty according to their separate generations" to use Mendel's words when he was setting out the reasons earlier experimentation had failed to establish the principles of hereditary transmission. And, indeed, as we have already seen when East and others, though dealing with quantitative variation which denied them the use of Mendel's type of analysis, nevertheless used his experimental structure of crossing true-breeding parents and raising F_1, F_2 *etc.*, they were able to gain clear evidence (albeit of a kind numerically different from Mendel's) of both segregation and recombination. It is experiments of this kind that have enabled us, following a lead given by Fisher, Immer and Tedin (1932) to devise modern methods of biometrical genetics (Mather, 1949; Mather and Jinks, 1971). These have allowed us not only to understand more fully the results obtained from the biometrical analysis of populations, but to investigate the properties in both linkage and action of the genes constituting the polygenic systems underlying quantitative variation -- though always, of course, with the limitation that we have already noted, *viz* that the information we obtain relates to the set of genes taken as a whole and not to individual gene-pairs. Furthermore, we gain the immense advantage of no longer being confined to the use of variances and co-variances, but in addition are able to compare the mean expressions of characters in parents, F_1, F_2, F_3 and other generations and gain valuable information from them. Using these means we can detect the presence of dominance and in suitable cases obtain estimates of the average level of dominance shown by the genes involved. We can see how the genes at different loci can balance as well as supplement one another's effects in the parents and how their dominance properties can be reinforcing or opposing. We can detect interaction between non-allelic genes (epistasis as it is now frequently called) and obtain information about the form which this takes.

The second degree statistics (variances and covariances) obtained from such experiments are also more informative than those from populations, partly because we can obtain so many more generations of known relationship to one another and partly because when we start off with a cross between true-breeding lines we start with equality of the gene frequencies, *i.e.*, $u = v = \frac{1}{2}$. This immediately reduces

$$D_R = 4\Sigma\{uv\left[d + (v-u)h\right]^2\}$$

to

$$D = \Sigma(d^2)$$

and

$$H_R = 16\Sigma\{u^2v^2h^2\}$$

to

$$H = \Sigma(h^2)$$

so giving us genetic components that depend on separate and distinct properties of the genes, each component representing the full contribution of that property to the variation. We can thus obtain, for example, an estimate of the average dominance ratio of the gene-pairs that are segregating. The effects of linkage and epistasis in changing the values of D and H can be specified, so enabling us to detect and measure in appropriate ways the consequences of these phenomena for the variation -- a test that is prohibitively difficult, even where it is theoretically possible, with data from populations. It is also possible by the introduction of appropriate parameters to detect and estimate the effects on the variation of any other phenomenon, such as sex-linkage or maternal effects, that we may think to be prospectively relevant (Mather and Jinks, 1971). In short, the use of experiments starting with deliberate crosses provides the possibility of investigating by biometrical means the properties of the genes underlying quantitative variation where direct Mendelian analysis cannot be used. We shall have occasion to refer later to some of the results from these analyses.

V. THE NUMBER OF GENES: DIRECT ESTIMATES

Having established that the heritable determinants of quantitative variation must be nuclear genes, the question immediately arises of how many genes are characteristically involved. Is it a large number or a small one? We have already had occasion to note that the observation of quantitative variation does not carry any necessary implication about the number of genes involved. Given that the non-heritable variation is large by comparison with the effects due to a single gene-difference, quantitative variation will ensue: and indeed quantitative variation is observable in homozygous lines where no gene differences at all are segregating, as Johannsen showed. Simple progeny testing can quickly reveal the absence of a genetical component in the variation and will commonly serve to pick out the cases where only a single gene difference is at work. Homozygous lines are, however, a special case seldom encountered outside naturally inbreeding species, except for lines that have been deliberately inbred. And cases of quantitative variation are not common that involve only so few gene differences that they can be sorted out unambiguously by simple progeny testing and test-crossing. More general means had to be devised for arriving at estimates of the number of genes at issue. The question may be approached either by direct analysis, where this is possible, or more generally by biometrical analysis.

The approach by direct analysis involves tracking down the genes one by one, and it therefore requires of each gene first that its effect is large enough for it to be traced, (using, of course, appropriate progeny testing where this is possible) and secondly that it can be distinguished from its fellow genes that are contributing to the variation. The first requirement sets a limit to the genes that can be traced: even with extensive progeny testing there will be a lower limit of effect below which a gene difference will not be detected. Thus the number of genes ascertained must always be regarded as the minimum number involved.

The second requirement, that the genes be distinguishable from one another, can in principle be met in either of two ways: by the genes being distinguishable through their action on the phenotype or through their locations in the chromosomes. As we shall see later, there is evidence that some of the genes in a system may have different initial actions and so contribute to the quantitative variation in different ways, but this does not imply that all of them in fact do so. Certainly there is little reason to expect that a polygenic system will be completely partitionable by this means in the forseeable future if ever at all. This leaves us with the second means of distinguishing the genes, by their locations in the chromosomes. In principle this could be done in any species by raising large numbers of inbred lines and testing them against one another by crossing and further progeny testing; but in practice, the task would become so demanding of facilities and time as to make it impracticable in all but simple cases of very small numbers of gene differences. In *Drosophila*, however, the availability of so many easily followed marker genes does make it possible to attempt to partition a polygenic system by finding the location of its constituent genes in the chromosomes and so to derive at least a minimal estimate of the number involved.

The technique of chromosome assay enabled us to distinguish among genes carried by the different chromosomes. To go further required the use of suitably marked tester chromosomes: recombinant chromosomes detected by the reassociation of the marker genes could then be examined for their genetical

activity in relation to the quantitatively varying character
and the distribution of this activity along the chromosome
worked out, with each active region implying at least one gene.
In its simplest form this technique enabled Wigan (1949) to
show that there were several regions of the X chromosome that
affected the number of abdominal chaetae, while Breese and
Mather (1957) used a more sophisticated form of this basic
technique to break chromosome III into six segments, each of
which showed activity for this same character and each of
which must, thus, be carrying at least one gene contributing
to its variation. The most powerful means of using recombination, detected by marker genes, to identify genes contributing to quantitative variation was, however, developed by
Thoday (1961) who showed that by this means, genes contributing to variation in sternopleural chaetae number could be
assigned with hitherto unattained precision to loci in the
chromosomes. This method has been used both extensively and
effectively by Thoday and his co-workers to analyse the
effects of various kinds of selection for sternopleural chaeta
number (see for example, Spickett and Thoday, 1966; Thoday,
1972). Davies and Workman (1971) showed that the genes mediating the quantitative variation of abdominal chaeta number
were not the same as those affecting the variation in sternopleural chaetae. And Davies (1971) used Thoday's method in
combination with results obtained by less precise techniques,
to show that at least some 14 or 15 genes must be involved in
both of these systems -- a finding which agrees well with the
estimate of about 18 genes affecting abdominal chaeta number
derived by Mather and Jinks (1971) in a more roundabout way
from earlier evidence. It should be noted, too, that these
estimates were each derived from experiments calling on a
limited range of ancestral material. So, unless we are pre-
pared to assume that despite this limitation genes at all the
relevant loci were detected, we must accept that the numbers
of genes that can, and over a wider range of material will,
contribute to the variation of these characters must be greater still. The variation can truly be described as polygenic.

VI. BIOMETRICAL ESTIMATES: EFFECTIVE FACTORS

Estimates of the number of genes contributing to quantitative variation have been attempted by several different
biometrical methods. They have been based on different types
of data drawn from different characters in different species,
and have yielded widely different results. "Student" (1934)

sought to estimate the number of genes involved in the responses obtained at the Illinois Experiment Station to selection for oil percentage in the seeds of a population of maize (see Winter, 1929). He compared the difference in the oil percentages of high and low lines after 28 generations of selection with the estimated genetical variance of the original population, and concluded that the number of genes involved was at least of the order 20-40, could possibly have been in the range 200-400, and was not at all likely to have been as low as 5-10. Falconer (1971) also drew on data yielded by selection lines from a population, his selection being for litter size in mice, but he used quite a different kind of calculation which involved consideration of the effects of inbreeding on the selection lines. He nevertheless also arrived at a high number, the estimate being that some 80 gene-differences must have been segregating in the base populations from which the selection started.

Much lower estimates have been obtained from the segregating generations obtained after crossing pairs of lines, estimates which in fact have seldom exceeded 7 or 8 and are commonly 4 or 5 or even lower (see Mather, 1949; Mather and Jinks, 1971). The oldest and most widely used method of estimating "k", the number of genes in the system, goes back to Wright (quoted by Castle, 1921) and finds k as the ratio of the square of the difference between the mean expressions of character in the parental lines to eight times the additive genetic variance of the F_2. It will always underestimate k if the effects of the different genes in the system are not equal, and also, of course, if the alleles mediating the greater expression are not all concentrated in one parent and the alleles mediating the lesser expression in the other. A second method of estimating k, based on the distribution of the genetic variances of F_3 families, was given by Panse (1940). This does not require that the alleles are co-directionally distributed between the parents, but gives estimates subject to even greater reductions by inequality of the individual gene effects. It also requires elaborate corrections for non-heritable variation and has been little used. A third method has recently beed developed by Jinks and Towey (1976) which depends on ascertaining the proportion of individuals heterozygous for at least one gene-difference in the F_2 or a later generation. It too will always yield a minimum estimate since gene differences of small effect may escape detection; but properly used it promises to be more useful than either of its predecessors.

All three methods give the low estimates of k that have already been noted. They all share the restriction of giving valid estimates only if the genes in the system are segregating independently of one another: whatever its phase, any linkage among the genes will reduce the estimate of k by an amount proportional to the intensity of the linkage. Genes lying near to one another in a chromosome will tend to appear in the same unit of segregation whose basis is, thus, not a single locus but a piece of chromosome. Such a unit has been termed an effective factor (Mather, 1949; and see Mather and Jinks, 1971). Estimates of k obtained by use of the segregating generations following a cross between two lines are, therefore, estimates not of the number of genes in the system but of the number of effective factors. It is of no avail to complain that estimates of the number of genes would be more useful: the number of effective factors is the only thing that can be estimated from data of this kind.

Since, however, genes which because of their linkage have segregated together at one round of meiosis may be separated by crossing over at the next, the effective factor is no more than a temporary unit and the number of them must be expected to increase as generation succeeds generation. This has in fact been found to happen by Towey and Jinks (1978) in *Nicotiana rustica*. What is more, the increase is rapid: the number of effective factors found by analysis of F_2 was no more than 3, but by F_6 it had risen to some 15 or more. This latter figure is in much better keeping with the estimates obtained by "Student" from maize and by Falconer from mice, in both of which cases the basis for the estimate came from material derived by many generations of open breeding and selection starting from a population. In other words, they came from material in which linkage would play a much less restrictive part in determining the effective size of the units of segregation and response to selection, and so would be expected to yield a higher estimate of k than would an F_2.

The great increase in the number of effective factors observed by Towey and Jinks raises another point. The increase is attributable to continuing recombination breaking up linkages between member genes of the system during the successive rounds of meiosis in the formation of the gametes from which succeeding generations were derived. Now *Nicotiana rustica* has 24 pairs of chromosomes. So, taking the larger number they observed, where say 20 genes were distributed at random over the total length of chromosome material, on average just over 10 of the 24 chromosomes would not be carrying at least one of the genes. So even if the genes were closely linked in any chromosome carrying more than one of them we should expect an estimate of something like 13, or slightly fewer, effective factors in F_2. With a total of 15 genes the estimate expected in F_2 would be some 8 or 9. Both expectations are much higher than the estimate of 3 obtained in F_2 by Towey and Jinks. One is, thus, led to the conclusion that the genes cannot have been scattered at random over the chromosomes, but must rather be clustered in a relative few chromosome regions: that, in fact, in respect of the quantitative variation that Towey and Jinks were observing there must have been a few active regions of the chromosome which were largely responsible for the variation. It is worth noting in this connection that there is some indication of clustering in the distribution of the genes mediating the variation of sternopleural and abdominal chaeta numbers in *Drosophila* (Davies, 1971).

The resolution of effective factors by their exposure to continuing recombination implies that new combinations of the genes can come into being and will be held together by the same linkages in the same pieces of chromosome as were the combinations from which they were derived. Where they were built up by rare recombination they will break down only by equally rare recombination: once in being they will be as persistent as they were difficult to produce. They will give new allelic forms of the effective factors so to speak; new units which, if selected for large effect on the character, may display a dominating influence in the control of its variation. In the lines of *Drosophila* selected by Thoday and Boam (1961) for high sternopleural chaeta number, Spickett and Thoday (1966) isolated three units which between them accounted for the bulk of the change in chaeta number from the ancestral level. One of these units was probably present in the original crosses from which the selection lines were derived. But it seems improbable that the other two were, or selection would not have taken as long as it did to bring about the advance ultimately achieved. The indication (Mather and Jinks, 1971) is clearly that they were assembled during the course of the selection by the recombination of closely linked elements which in the ancestral chromosomes were present in different combinations having lesser effects on chaeta number -- an interpretation to which Thoday (1973) also subscribes and to whose implications for genic evolution he also draws attention.

VII. GENE ACTION

The genes of continuous variation are borne by the chromosomes just like the major genes of Mendelian genetics. They are, however, commonly found to be acting in polygenic systems and the properties of quantitative variation lead us to postulate the member genes of the system as having small, similar and supplementary effects, in apparently striking contrast to the drastic and specific effects of the major genes. How far is this apparent contrast valid? Does it arise from a fundamental difference in action between two distinct classes of genic structure? Or is it artificial, reflecting no more than a difference in approach, quantitative genetics being concerned with all the genes contributing to a particular kind of variation in a particular character, whereas major genes are picked out no matter what character they affect because the magnitude and constancy of their effects make them easy to recognise and follow in genetic studies. Before considering these questions in more detail let us first look at the information we have about the action of polygenic systems.

As yet molecular genetics offers little to help us, but we have a growing amount of information obtained in other ways. First of all the observations of Spickett and Thoday (1966) leave no doubt that gene differences contributing to quantitative variation can differ from one to another in the magnitudes of their effects. At the other end of the genic scale, allelic series like that at the white locus of *Drosophila* show that major genic structures can be associated with wide variation in the size of effect of their allelic differences, the differences of effect being sometimes so small as to require special techniques for their detection (Muller, 1935). Thus in respect of size of effect, members of a polygenic system and major gene differences may well represent the extremes of a continuous spectrum of magnitudes -- though whether the gene differences in the centre of the spectrum are as common as those at the extremes is by no means clear.

Turning to other properties of gene action, Fisher (1918) showed that the genes mediating continuous variation in man displayed at least partial dominance. This finding has been confirmed and extended by innumerable biometrical studies on both plants and animals since that time. Similarly in more recent times the interaction of non-allelic genes has been demonstrated and measured by biometrical methods (see Mather and Jinks, 1971). One cautionary point must, however, be made. Both the degree and even the direction of dominance and also of non-allelic interaction can be altered by the choice of scale on which the character is measured; but the use of an unfortunate scale cannot account for all or even most of the cases that have been observed. There can in fact be no doubt of either dominance or non-allelic interaction as a feature of genes in producing quantitative variation, just as they are a commonplace of major genes. In quantitative variation dominance and interaction would, however, appear commonly to be partial rather than complete as they so frequently are with major genes.

Where the use of appropriate techniques allows us to recognise individual gene differences contributing to quantitative variation it is possible to investigate not only their properties of dominance and interaction but also the way they act during development. In an early study Spickett (1963) was able to show that one gene difference affecting sternopleural chaeta number did so by increasing cell number whereas another did so by affecting the time of initiation of one of the macrochaetae. Thus at least some of the genes contributing to continuous variation do so in specifically different ways. To this extent it has become clear that the similarity in effect of the members of a polygenic system may be superficial in that when we analyse the ways in which they contribute to the variation we find that the similarity vanishes (see also the chapter by C.E. Caten in this book). But whether such specific differences of action would be found between all the members of a polygenic system is another question. The gene differences so far analysed in this way are necessarily those having relatively large effects and to say the least it would be unwise to assume that all the genes of lesser effect share the same property. Indeed in discussing his findings, Thoday (1973) points out that the genes Spickett studied were the units built up presumably by recombination during Thoday and Boam's (1961) selection experiment. He goes on to note that these results suggest that material of rather unspecific effect can be brought together by recombination to produce genes of more specific effect and raises the question of the extent to which such a process might be involved in gene evolution. It can hardly be doubted that genes must have evolved as units of action, and this process of combining elements of unspecific effect into units at once more complex and more specific is one that could be intimately involved in the production of the genes that genetical studies have revealed to us (Mather, 1949, 1954).

One last point remains to be noted about the action of genes mediating quantitative variation. Studies of such variation in the number of chaetae borne on the coxae of the three pairs of legs of *Drosophila* indicate that the relevant genes fall into three classes: α genes acting equally on the coxae of all legs; β genes which are inactive on the rear but active equally on the coxae of middle and front legs; and γ genes active only on the coxae of the front legs (Mather and Hanks, 1978; Hanks and Mather, 1978). While other explanations cannot fully be ruled out, the most attractive interpretation of these results is that each of the genes has a controlling element determining in which legs its structural element will be active in affecting chaeta production. This interpretation implies, of course, still another (though perhaps not unexpected) similarity between major genes and the members of polygenic systems mediating quantitative variation. It also implies yet another way in which genes producing similar effects might differ from one another, and yet another kind of change in the genetic material that can affect quantitative variation.

VIII. THE GENIC STRUCTURES INVOLVED

The evidence from gene action gives little indication of any fundamental difference in the genetic structures responsible for quantitative variation on the one hand and for major genic differences on the other. The relations between the two types of variation need not, however, always be the same. They may in fact be of three different kinds (Mather, 1949, 1954) *viz*:

(i) The same difference in the same physical structure might produce a phenotypic difference from which a major gene would be inferred while simultaneously contributing, together with other genic differences, to quantitative variation.

(ii) The same structure might be capable of varying in two separate ways on separate occasions, one leading to the inference of a major gene and the other to the inference of a member of a polygenic system (or polygene as it has been called).

(iii) Distinct kinds of structure might exist, differences in one kind leading to major genic effects and in the other to polygenic effects.

The first relationship, which is one of pleiotropy, has been emphasized by Penrose (1951) and Grüneberg (1952), and Penrose cites the case of the gene for phenylketonuria in man which is unquestionably to be classed as a major gene in respect of its effect on mental capacity but which contributes to quantitative variation in other characters. That gene changes having a major effect on one character should have minor effects on others is perhaps to be expected: pleiotropic effects are common and their developmental ramifications complex. To ascribe any sizeable proportion of the ubiquitous quantitative variation that we see around us to such pleiotropic relations as Penrose was discussing is, however, unrealistic, for the major differences necessary on this view are not to be found. Nor can this be reasonably attributed to their being cryptic, affecting hidden characters.

Indeed, until they have been uncovered and demonstrated in at least one case, this postulate of hidden characters cannot be given serious credence.

In recent years an interesting variant of this view has been put forward, relating quantitative variation to the polymorphisms for genically controlled isozyme variation found so commonly in natural populations. Certainly some of the variation in some characters is to be attributed to such polymorphism. Thus Gibson (1970) and Birley and Barnes (1973) have shown that the polymorphism for fast and slow alleles of the alcohol dehydrogenase gene (Adh) in *Drosophila* is a major contributor to the variation in the level of activity of the enzyme produced; but even so it still leaves a substantial part of the variation to be accounted for, and most of it is traceable to the activity of genes on chromosome III, as distinct from chromosome II on which the Adh gene itself is located (Barnes and Birley, 1978). Again of course, it might be postulated that other enzyme polymorphisms yet to be detected are also involved, but until at least one case has been analysed and shown to conform to this interpretation we can at best do no more than reserve judgement.

The third relationship lies at the other extreme from the first, in involving distinct differences in distinct structures. The possibility of this relationship was sharpened by two pieces of evidence that came available in the 1940's. It has long been known that certain species of both plants and animals can carry two types of chromosomes, the basic complement and the so-called B chromosomes (see Darlington, 1937). These latter vary in number from one individual to another, some carrying no B chromosomes at all. Thus unlike the basic complement, they are not essential for the successful development of the organism and their variation in number does not have the drastic effect on the phenotype well known to arise if whole chromosomes or pieces of them from the basic complement are lacking or present in excess. They clearly differ in their genetical activity from the chromosomes of the basic complement, and indeed have often been described as inert. They also differ cytologically in showing the property of being heterochromatic. In 1943 Müntzing reported that, despite their apparent inertness, they did affect, albeit in a relatively small way, the nuclear processes in *Sorghum* and morphology and fertility in rye, which characteristically show quantitative variation. Many observations have confirmed that B chromosomes occur in many species of plants and animals and that they exert their effects on a great variety of characters, including at least certain aspects of the behaviour of the basic complement of chromosomes (Rees and Jones, 1977).

The Y chromosome of *Drosophila* shares many of the properties of B chromosomes. It is heterochromatic; it is not essential for development in males, where it normally occurs (though it is essential for the production of motile sperm); and it can vary in number in both sexes without any consequences comparable with those of unbalance even for the small chromosome IV. Apart from *bobbed*, it has not shown any evidence of genes capable of giving major changes of the phenotype. Yet evidence has been reported that it could contribute to quantitative variation in chaeta number (Mather, 1944) and cell size (Barigozzi, 1951). Thus once again heterochromatin, while failing to reveal any indication of carrying genes essential for adequate development of the organism, showed polygenic activity, as pointed out by Mather (1944). Evidently genes physically different from the major genic structures of euchromatin could contribute to quantitative variation. This, of course, left open the question of whether quantitative variation sprang wholly from the action of such genes, as was indeed recognised at the time by Mather's statement "Thus Payne's case is not conclusive evidence of the polygenic activity of euchromatin; but it would be unwise as yet to go to the other extreme and make the assumption, attractive though it is, that euchromatin is devoid of polygenes. A detailed survey of some chromosome or chromosomes will first be necessary." (Despite this, it has frequently been said, even in recent years, that he attributed all genetically determined quantitative variation to heterochromatin and hence to a physically distinct and separate type of gene.)

A survey was not long in coming. It was shown by Wigan (1949) that polygenic activity on the X chromosome of *Drosophila* as displayed by variation in abdominal chaeta number, could not wholly be accounted for by heterochromatin and that much of it must be traced to euchromatic parts of the chromosome, thus confirming the indications afforded by Payne's (1918) earlier and less certain results. Wigan's findings have been fully substantiated and extended by later work on the location of the genes affecting chaeta number (Davies, 1971).

This leaves us with the second relationship to consider, that the same genic structure might vary in two distinct ways, one leading to the inference of a major gene and the other to that of a member of a polygenic system. The latter type of change would, thus, lead to the production of iso-alleles, effectively alike in respect of the major genic activity but differing in their effects on the quantitative variation. The first case of iso-alleles to be reported was at the *ebony* locus in *Drosophila melanogaster* (Stern, 1926), but perhaps the most interesting of the early cases was that of two wild-type alleles of *white* differing in their mutation rates when irradiated (Timoféeff-Ressovsky, 1932) and later shown by Muller (1935) to have slightly different properties of dominance in triploid females. Both of these cases involve "wild-type" alleles giving phenotypically indistinguishable homozygotes but separable by their dominance properties. They leave little doubt that iso-alleles of the kind in which we are interested can arise.

The relations observed by Reeve and Robertson (1953) between lethality and quantitative variation in certain body size characters led them to propose an interpretation in terms of iso-alleles, but their conclusion is by no means certain. It is very difficult indeed to be confident that one is dealing with changes in a structure unitary in its major genic effect but sufficiently complex to produce iso-alleles differing in their quantitative effects, as distinct from an effective factor based on a genetically very short piece of chromosome but carrying separate structures displaying the major genic and polygenic effects respectively (Mather, 1954). To do so, indeed, requires some criterion for a unitary genic structure, which leads us into considerations of cis-trans effects and complementation, and which in doing so would take us beyond our present limits. Two points may, however, be made.

In the first place, perhaps we could and should regard Birley and Barnes' observations on the alcohol dehydrogenase gene as such a case, the fast and slow forms being iso-alleles in respect of their effects on the overall activity of the enzyme, but iso-alleles that are readily distinguishable because of the special power of the electrophoretic technique by which they are separated.

The second point is that there is a good theoretical case for considering that iso-alleles can contribute to quantitative variation. Indeed some views on gene evolution (Mather, 1949, 1954; Thoday, 1973) would lead us to expect the occurrence of such iso-alleles. What proportion of quantitative variation might be ascribed to them is impossible to say at present. We have, however, come to the stage where we can recognise that genic structures of more than one kind can contribute to polygenic systems, and that perhaps, as Thoday has suggested, "any kind of locus could contribute to polygenic variation."

IX. THE IMPACT OF SELECTION

The impact of selection on quantitative variation has been studied in two distinct, though by no means wholly separable, connections. First, it has been looked at in relation to the changes it brought about in the expression of the characters upon which it was being practised. This approach goes back to well before 1900, when the significance of Mendel's findings came to be appreciated; and it has commanded great interest from breeders concerned with the improvement of our domesticated plants and animals, which characteristically display quantitative variation in virtually all of their productively important characters. The second and more fundamental point of view stems from Darlington's (1939) pioneering book "The Evolution of Genetic Systems", of which the theme is that selection will adjust not only the expressions of characters themselves but also the organisation and properties of the genetic materials underlying the variation and changes in the manifestation of those characters. Darlington was concerned primarily with the chromosomes and their properties, but the concept of the genetic system as he called it has been extended to include the breeding system of the species, whether inbreeding or outbreeding, and in our immediate context, the organisation of polygenic systems.

The first major selection experiment was begun at the Illinois Experiment Station in 1896 and was for the oil and protein content of maize (corn) grains (Smith, 1908). It was continued for several decades and the results from the first 28 years were reported by Winter (1929). The changes under selection were so great that "Student" (1934) commented that in both high and low selection lines the levels of the oil content "were clean outside the original range". It was in fact the first experimental revelation of the amount of variation that can lie hidden in balanced polygenic combinations within a population. Similar results have been obtained from many subsequent experiments with other species, an interesting early example being Payne's (1918) selection for the number of bristles in *Drosophila*.

The early experiments showed what selection could do, but a more analytical approach was necessary if the capacity of a population for change under selection was to be measured, as breeders would obviously wish it to be, and efficient modes of selection devised to achieve the desired change. Fisher's (1918) partition of variation provided the necessary basis. It was then a simple step to the notion of heritability (often incorrectly, and misleadingly, attributed to the character rather than to the variation). Since, as Fisher showed, the

genetic variability can be of different kinds stemming from
different genetic effects, different measures of heritability
are obtained according to which kinds of genetic variability
are included in the fraction, ranging from the additive component alone (leading to the "narrow" heritability) to the totality of the genetic components (leading to the "broad" heritability). On the basis of the heritability, denoted by h^2,
and the degree of inbreeding in the mating system, calculated
from the ancestry of the individuals involved and denoted by
F (Wright, 1921), a biometrical approach to estimating, predicting, and maximising responses of populations to selection
was elaborated between the wars. This was done chiefly by
animal breeders, who are necessarily concerned with populations, since the maize breeders had by this time become almost
exclusively involved with hybrid breeding, and the other plant
breeders were commonly dealing with inbreeding species and in
any case almost always started with crosses between pairs of
lines and so needed different biometrical methodology (see
Professor J.L. Jinks in this volume).

The progress of the biometrical application of genetics to
animal breeding in these early years was marked by notable
books from Lush (himself the most significant of its founding
fathers) in 1937 and Lerner in 1950. It was further developed
in the 1950's and later years, particularly by the use of
experimental populations of laboratory animals, notably mice
and *Drosophila*, for testing and elaborating the basic genetical concepts (see Falconer, 1960). Selection experiments
with *Drosophila*, carried out by Professor Alan Robertson and
his colleagues, have been especially extensive and informative
(Clayton *et al.*, 1957; McBride and Robertson, 1963; Sen and
Robertson, 1964; Robertson, 1966). Though hybrid breeding
methods, stemming from those developed for maize, are now the
basis for poultry improvement, selection in the larger stock
mammals has continued to be firmly based on this biometrical
approach, including the elaborate breeding schemes developed
to take advantage of artificial insemination as a tool in
raising milk production in cattle (Rendel and Robertson, 1950;
Robertson and Rendel, 1950).

X. ORGANISATION OF POLYGENIC SYSTEMS

Quantitative variation is a feature of virtually all characters in all species, wild as well as domesticated, as innumerable observations have revealed: it has been found wherever
it has been sought. We must expect, therefore, following

Darlington (1939), that there has been ample opportunity for selection, both natural and artificial where this has been applied, to have moulded the genetic system or system of variation (as it may perhaps more aptly be termed in our present context) in ways which will reveal themselves when further selection is applied.

The first experimental evidence of the effect of the system of variation on response to selection is afforded by comparison of the results obtained from the Illinois Experiment with maize and Johannsen's selection in beans. In maize, which is an outbreeder, steady and ultimately dramatic responses to the selection were obtained over decades. In the bean, which inbreeds by self-pollination, selection was without effect once the basic genotypes were sorted out in the first generation. The breeding system, whether outbreeding or inbreeding, is thus a key feature of the system of variation. This is, of course, to be expected from Mendel's principles and it can hardly have come as a surprise to most plant-breeders who were familiar with the use of single-ear selection, though it appears to have escaped De Vries (1907).

Another contrast between naturally outbreeding and inbreeding species that must early have become clear to plant breeders is that whereas, when artificially imposed on outbreeders, inbreeding results in inbreeding depression (manifesting itself in poor growth, lack of vigour and infertility) which can be immediately reversed by intercrossing plants to produce hybrid-vigour or heterosis, the naturally inbred individuals of an inbreeding species are not weak and infertile and they do not show any striking heterosis when intercrossed, even where the cross is between individuals of manifestly different genotypes. Thus, dominance, to which heterosis and its reverse expression, inbreeding depression, were early attributed (see Keeble and Pellew, 1910; Jones, 1917), must be differently adjusted under the two breeding systems. That adjustments of balance are produced by natural selection was made clear by Dobzhansky's (1948, 1950) finding that in *Drosophila pseudoobscura* homologous polygenic combinations were co-adapted to produce high fitness when they came from the same population where they would be selected to work together, but not when they came from different populations, where they would not. Thus, the homologous combinations that we find within a population are those which because of their relational balance or co-adaptation natural selection has picked out from among the wider range of combinations that can and do arise. In inbreeding species the polygenic combinations must nearly always be in the homozygous state and by the same principle must, therefore, have been selected from a wider range of combinations to give adequate vigour and fertility when

homozygous, whereas in outbreeders where such homozygotes will
be rare the combinations will not have been so selected and,
hence, will not in general be adequately balanced within themselves, and so will commonly lead to inbreeding depression
when made homozygous by artificial selection. Thus, the past
action of selection shows its effects in the present system of
variation, which in its turn will be a major determining factor of response to the future action of selection.

The action of selection in favouring certain balanced combinations of genes must imply the occurrence in populations of
these combinations at higher frequencies than random association of the genes in question would yield (Mather, 1941).
Such linkage disequilibrium, as it has come to be called,
though doubted in the early days (Wright, 1944), has now been
recorded in inbreeding species of plants (Allard et al., 1972;
Clegg et al., 1972) and in *Drosophila melanogaster* (Birley,
1974), which is, of course, an outbreeder. Furthermore, the
disequilibria are much greater in the inbreeder than the outbreeder and extend to combinations of unlinked genes. As
Allard et al. point out, the key feature in producing the equilibria is the restriction of recombination between the relevant genes, and this can arise in a more comprehensive way by
the restriction of crossing than by linkage (even where the
linkage is tightened by inversion heterozygosity), thus illus-
trating the complexity of the interrelation between breeding
systems, chromosome behaviour and natural selection in mould-
ing the system of variation.

The nature of the selection itself plays its part. Three
basic component types of selection can be recognised (Mather,
1953a). When a single optimum phenotype is favoured, selection towards it may be stabilising where the optimum is a
central phenotype near the mean of the population, or it may
be directional where the optimum is a more extreme phenotype.
Fisher (1928) pointed out that dominance will be built up by
selection, the favoured allele becoming dominant over its less
favoured fellows. Thus with aspects of the phenotype, like
so-called fitness characters, that must be under preponderantly directional selection, dominance should be preponderantly
in the direction of increased manifestation of the character.
With stabilising selection towards a central optimum, however,
selection on some individuals will be in one direction but on
others it will be in the other. In such cases dominance will
not be unidirectional but ambidirectional, and in crosses involving many gene differences it may even appear to be absent
because of the cancelling effects of the opposing dominances
of the individual genes (Mather, 1960). This expectation has
been verified in *Drosophila* (Breese and Mather, 1960; Kearsey

and Kojima, 1967), and indeed the type of dominance, uni- or ambi-directional is now frequently taken as evidence of the nature of the past selection.

The third basic type of selection arises where there is more than one optimum phenotype and is termed disruptive. The effect on the organisation will then depend on the relations between the individuals being selected towards the different optima. Where these individuals are independent of one another in their properties of reproduction and fitness there will be a tendency towards the rise of sub-populations showing increasing genetic isolation from each other. Where, on the other hand, the individuals are dependent on one another in respect of reproduction or fitness, as with males and females in a dioecious species or with the various morphs in species displaying Batesian mimicry, polymorphism will ensue (Mather, 1955). This expectation too has been borne out, and much of the genetic mechanism involved has been revealed, in the extensive and informative series of experiments carried out by Professor Thoday and his colleagues (reviewed by Thoday, 1972).

Finally we come to adjustment of the breeding system itself. Though we have no direct evidence of change under natural selection, we know both that variation exists in the wild for the degree of inbreeding and that this character can be changed by artificial selection in experiment (Breese, 1959). We may, therefore, reasonably infer that it has been adjusted in the wild by natural selection arising from temporal and spatial variation of the environment. We are, thus, led to see all the basic features of the organisation of polygenic variation as interlocked with one another both in their origins through past selection and in the determination of responses to current selective forces, whether natural or imposed artificially in experiment.

Our discussions of these interrelations has necessarily been cursory: to have done more would have made prohibitive demands on space. A fuller general treatment can be found elsewhere (Mather, 1973) and certain aspects are dealt with in more detail in later sections of this book. Like other facets of the study of quantitative variation, its organisation and the factors that determine this organisation are commanding growing attention -- as indeed they must if we are to understand how living populations have come to be as they are, the agencies that will govern their changes in the future and, in the case of domesticated species, the ways that we can manipulate them for the purposes of increased production.

REFERENCES

Allard, R.W., Babbel, G.R., Clegg, M.T., and Kahler, A.L. (1972). *Proc. Nat. Acad. Sci., U.S.A., 69,* 3043-3048.
Barigozzi, C. (1951). *Heredity, 5,* 415-432.
Barnes, B.W.,and Birley, A.J. (1978). *Heredity, 40,* 51-57.
Birley, A.J. (1974). *Heredity, 32,* 122-127.
Birley, A.J., and Barnes, B.W. (1973). *Heredity, 31,* 413-416.
Breese, E.L. (1959). *Ann. Bot., 23,* 331-344.
Breese, E.L., and Mather, K. (1957). *Heredity, 11,* 373-395.
Breese, E.L., and Mather, K. (1960). *Heredity, 14,* 375-399.
Castle, W.E. (1921). *Science, N.S., 54,* 233.
Clayton, G.A., Robertson, A., and Others. (1957). *J. Genet., 55,* 131-180.
Clegg, M.T., Allard, R.W., and Kahler, A.L. (1972). *Proc. Nat. Acad. Sci., U.S.A., 69,* 2474-2478.
Cooke, P., and Mather, K. (1962). *Heredity, 17,* 211-236.
Darlington, C.D. (1937). "Recent Advances in Cytology" (2nd edition). Churchill, London.
Darlington, C.D. (1939). "The Evolution of Genetic Systems." Cambridge University Press, Cambridge.
Davies, R.W. (1971). *Genetics, 69,* 363-375.
Davies, R.W., and Workman, P.L. (1971). *Genetics, 69,* 353-361.
Dobzhansky, T. (1948). *Genetics, 33,* 588-602.
Dobzhansky, T. (1950). *Genetics, 35,* 288-302.
East, E.M. (1915). *Genetics, 1,* 164-176.
Emerson, R.A., and East, E.M. (1913). *Bull. Agr. Exp. Sta. Nebraska, Res. Bull., 2.*
'Espinasse, P.G. (1942). *Nature, 149,* 732.
Falconer, D.S. (1960). "Introduction to Quantitative Genetics." Oliver and Boyd, Edinburgh.
Falconer, D.S. (1971). *Genet. Res., 17,* 215-235.
Fisher, R.A. (1918). *Trans. Roy. Soc. Edin., 52,* 399-433.
Fisher, R.A. (1928). *Amer. Nat., 62,* 79-92.
Fisher, R.A., Immer, F.R., and Tedin, O. (1932). *Genetics, 17,* 107-124.
Galton, F. (1889). "Natural Inheritance." Macmillan, London.
Gibson, J. (1970). *Nature, 227,* 959-960.
Grüneberg, H. (1952). *J. Genet., 51,* 95-114.
Hanks, M.J., and Mather, K. (1978). *Proc. Roy. Soc., B, 202,* 211-230.
Jinks, J.L., and Towey, P. (1976). *Heredity, 37,* 69-81.
Johannsen, W. (1909). "Elemente der exakten Erblichkeitslehre." Fischer, Jena.
Jones, D.F. (1917). *Genetics, 2,* 466-479.

Kearsey, M.J., and Kojima, K. (1967). *Genetics, 56*, 23-37.
Keeble, F., and Pellew, C. (1910). *J. Genet. 1*, 47-56.
Law, C.N. (1967). *Genetics, 56*, 445-461.
Lerner, I.M. (1950). "Population Genetics and Animal Improvement." University Press, Cambridge.
Lush, J.L. (1938, 1943, 1945). "Animal Breeding Plans," three editions. Collegiate Press, Ames, Iowa.
McBride, G., and Robertson, A. (1963). *Genet. Res., 4*, 356-369.
Mather, K. (1941). *J. Genet., 41*, 159-193.
Mather, K. (1942). *J. Genet., 43*, 309-336.
Mather, K. (1944). *Proc. Roy. Soc., B, 132*, 308-332.
Mather, K. (1949). "Biometrical Genetics," 1st edition. Methuen, London.
Mather, K. (1953a). *Symp. Soc. Exp. Biol., 7*, 66-95.
Mather, K. (1953b). *Heredity, 7*, 297-336.
Mather, K. (1954). *Proc. XIth Int. Cong. Genetics, Caryologia* Suppl. Vol., 106-123.
Mather, K. (1955). *Evolution, 9*, 52-61.
Mather, K. (1960). *Evoluzione e Genetica*, 131-152 Academia Nazionale dei Lincei, Rome.
Mather, K. (1973). "Genetical Structure of Populations." Chapman and Hall, London.
Mather, K., and Hanks, M.J. (1978). *Heredity, 40*, 71-96.
Mather, K., and Harrison, B.J. (1949). *Heredity, 3*, 1-52 and 131-162.
Mather, K., and Jinks, J.L. (1971). "Biometrical Genetics," 2nd edition. Chapman and Hall, London.
Muller, H.J. (1935). *J. Genet., 30*, 407-414.
Müntzing, A. (1943). *Hereditas, 29*, 91-112.
Nilsson-Ehle, H. (1909). "Kreuzunguntersuchungen an Hafer und Weizen." Lund.
Payne, F. (1918). *Indiana Univ. Stud., 5*, 1-45.
Panse, V.G. (1940). *Ann. Eugenics, 10*, 76-105.
Penrose, L.S. (1951). *Ann. Eugenics, 16*, 134-141.
Punnett, R.C. (1950). *Heredity, 4*, 1-10.
Rasmusson, J.M. (1935). *Hereditas, 20*, 161-180.
Rees, H., and Jones, R.N. (1977). "Chromosome Genetics." Arnold, London.
Reeve, E.C.R., and Robertson, F.W. (1953). *J. Genet., 51*, 276-316.
Rendel, J.M., and Robertson, A. (1950). *J. Genet., 50*, 1-8.
Robertson, A. (1966). *Proc. Roy. Soc., B, 164*, 341-349.
Robertson, A., and Rendel, J.M. (1950). *J. Genet., 50*, 21-31.
Sax, K. (1923). *Genetics, 8*, 552-560.
Sen, B.K., and Robertson, A. (1964). *Genetics, 50*, 199-209.
Smith, L.H. (1908). *Ill. Agr. Exp. Sta. Bull., 128*, 459-575.

Spickett, S.G. (1963). *Nature, 199,* 870-873.
Spickett, S.G., and Thoday, J.M. (1966). *Genet. Res., 7,* 96-121.
Stern, C. (1926). *Z.I.A.V., 41,* 198-215.
"Student" (1934). *Ann. Eugenics, 6,* 77-82.
Thoday, J.M. (1961). *Nature, 191,* 368-370.
Thoday, J.M. (1972). *Proc. Roy. Soc., B, 182,* 109-143.
Thoday, J.M. (1973). *Atti del. Accad. del. Sci. Bologna. Mem. Ser. III,* No. 1, 15-25.
Thoday, J.M., and Boam, T.B. (1961). *Genet. Res., 2,* 161-176.
Timoféëff-Ressovsky, N.W. (1932). *Proc. 6th Int. Cong. Genetics, 1,* 308-330.
Towey, P., and Jinks, J.L. (1978). *Heredity, 39,* 399-410.
De Vries, H. (1907). "Plant Breeding." Chicago.
Warren, D.C. (1924). *Genetics, 9,* 41-69.
Wigan, L.G. (1949). *Heredity, 3,* 53-66.
Winter, F.L. (1929). *J. Agric. Res., 39,* 451-475.
Wright, S. (1921). *Genetics, 6,* 111-178.
Wright, S. (1944). *Ecology, 26,* 415-419.

QUANTITATIVE GENETIC VARIATION IN FUNGI

C. E. Caten

Department of Genetics
University of Birmingham
Birmingham, U.K.

I. INTRODUCTION

A. *Fungi as Model Systems in Genetics*

The fungi possess many biological and technical properties which make them highly suitable organisms for genetical studies. These include: short life cycles, rapid growth, a variety of genetic systems (both meiotic and mitotic), a dominant haploid phase, ready mutant induction and isolation, ability to be cloned and growth on defined media under controlled laboratory conditions. Extensive exploitation of these features in a few species has contributed greatly to our knowledge of fundamental genetic and cellular mechanisms (see Esser and Kuenen, 1967; Fincham and Day, 1971; Burnett, 1975, for reviews of Fungal Genetics). In contrast, fungi have been little used for investigations of quantitative variation despite the fact that all the technical advantages of these organisms apply equally to the analysis of continuous variables. Consequently fungi have contributed little to our understanding of the genetic control of continuous variation and quantitative inheritance is considered in only one of the above mentioned texts (Burnett, 1975). This chapter aims to describe the methods of analysis of quantitative variation that have been developed in fungi, to summarise what these analyses have revealed, and to illustrate the potential of this group of organisms for studies of the nature and physiology of polygenic systems.

B. *Sources of Quantitative Genetic Variation*

Quantitative variation is typically encountered wherever a measurement character is influenced by several genes segregating simultaneously and is sensitive to environment or other non-heritable effects. In general, fungal geneticists have avoided this situation by: (1) working with closely-related strains differing at only a few known mutant loci, (2) concentrating on characters close to the primary action of the genes which are more likely to vary in a discrete manner and to be less sensitive to environmental effects, and (3) retaining only mutations with a large phenotypic effect and discarding those which are difficult to classify qualitatively. This approach has been highly successful in revealing the genetic systems of the organisms and in determining the location, fine structure, function, and regulation of individual genes (Esser and Kuenen, 1967; Fincham and Day, 1971). However, where mycologists have chosen or have been forced to work with unrelated strains, to consider complex continuously varying characters and to retain all variation irrespective of the magnitude of its effect, the existence and importance of quantitative genetic variation has become apparent.

The natural variability of fungal pathogen populations is critical in determining the stability of host resistance and fungicide effectiveness (Day, 1974). Comparison of independent isolates from natural populations reveals that many characters of pathogenic and ecological importance vary in a continuous manner (Holt and Macdonald, 1968; Brasier, 1970; Schwarzbach and Wolfe, 1975; Croft and Jinks, 1977). Quantitative variation is not restricted to natural isolates; it also arises through mutagenic treatment in the laboratory. For a measurement character, such as penicillin production, the effect of mutagenic treatment is to increase the variance between sister isolates rather than to produce discrete mutants (Simpson and Caten, 1979). The yield of most microbial products varies in this way and industrial strain improvement involves the stepwise accumulation through recurrent mutagenesis of favourable mutations, frequently of individually small effect (Alikhanian, 1970; Elander and Espenshade, 1976). As a result, independent highly-productive strains are separated by many mutational steps and hybridization would be expected to release a continuous range of productivities. Thus, a need for methods of handling quantitative genetic variation in fungi arises both from involvement with natural populations and from breeding for industrial purposes. In addition to these applications fungi may, because of their particular technical properties, contribute to our basic knowledge of quantitative genetic variation.

C. *The Systems*

Quantitative genetic variation has been examined in only a few species of fungi, with the most extensive studies involving the Ascomycetes *Neurospora crassa* and *Aspergillus nidulans* and the Basidiomycete *Schizophyllum commune*. A variety of characters has been examined, some selected for their technical convenience, *e.g.* colony radial growth rate and fruiting time, and others for their ecological or industrial importance, *e.g.* pathogenic aggressiveness and penicillin production. The methods developed for the analysis of continuous genetic variation in fungi and the results obtained will be illustrated by the most extensively studied systems, with reference to other systems to illustrate the range of application, to support any generalizations, and to point out where these have given results different from those of the model systems.

II. ANALYSIS OF QUANTITATIVE GENETIC VARIATION IN FUNGI

In most fungi the dominant phase in the life cycle is haploid and an independent sexual diploid phase is restricted to some yeasts, Myxomycetes and Oomycetes (Raper, 1966; Caten and Day, 1977). Basidiomycetes, however, possess an independent dikaryotic phase which genetically resembles the diploid state (Simchen and Jinks, 1964). These two ploidies, haploid and diploid/dikaryotic, require different methods of quantitative analysis; those used for the latter are essentially similar to the procedures developed for higher organisms while those used for haploids involve considerable modifications (Mather and Jinks, 1971).

A. *Haploids*

1. *Bi-parental Crosses*

a. Random spores. Considerable information can be obtained from a simple cross involving two parents and a sample of their F_1 progeny. (In haploids the F_1 is a segregating generation.) Two or more replicate clones of each of the parents and F_1 progeny are compared in a single randomised experiment. A simple analysis of variance on the progeny data (Table I) provides a test for the presence of genetic differences and estimates of the environmental (σ_E^2) and genetic (σ_G^2) components of variation. A significant between progeny

Table I. Analysis of variance for the F_1 progeny from a bi-parental cross in a haploid.

Source of Variation	Degrees of Freedom[a]	Expected Mean Square[b]
Between F_1 progeny	$s - 1$	$\sigma_E^2 + n\,\sigma_G^2$
Between clones within progeny	$s(n - 1)$	σ_E^2

[a] s = number of progeny; n = number of replicate clones.

[b] σ_E^2 = environmental component of variation; σ_G^2 = genetic component of variation.

mean square indicates segregation among the F_1 and, hence, that the two parents were genetically different. Given a genetic difference between the parents, their means are defined as follows:

$$\overline{P}_1 = m + [d] + [i] \quad \text{and} \quad \overline{P}_2 = m - [d] + [i],$$

where m is the overall mean, $[d]$ is the contribution of additive effects, and $[i]$ is the contribution of non-additive effects (Butcher, 1969; Mather and Jinks, 1971). The F_1 progeny segregate 1:1 for the alleles at each locus and hence the additive contributions of each allele pair to the F_1 progeny mean cancel. Likewise, in the absence of linkage, all possible genotypes will be produced in equal frequency and the interactions between genes (the sole cause of non-additive effects in haploids) will also cancel. Hence, the F_1 is expected to have a mean of m ($\overline{F}_1 = m$) (Butcher, 1969; Mather and Jinks, 1971). The parameters m, $[d]$ and $[i]$ and their variances (V) may then be estimated as follows (Mather and Jinks, 1971):

$$m = \overline{F}_1 \qquad\qquad V_m = V_{\overline{F}_1}$$
$$[d] = \tfrac{1}{2}(\overline{P}_1 - \overline{P}_2) \qquad V[d] = \tfrac{1}{4}V_{\overline{P}_1} + \tfrac{1}{4}V_{\overline{P}_2}$$
$$[i] = \tfrac{1}{2}(\overline{P}_1 + \overline{P}_2) - \overline{F}_1 \qquad V[i] = V_{\overline{F}_1} + \tfrac{1}{4}V_{\overline{P}_1} + \tfrac{1}{4}V_{\overline{P}_2}$$

Comparison of the mid-parental and F_1 means provides a simple test for the presence of non-additive variation. Where a model involving additive genetic and environmental effects

TABLE II. Genetic variation in bi-parental crosses in haploids (Adapted from Caten and Jinks, 1976).

Species	Character	Number of Crosses		Heritability[a]		Reference[b]
		Total	$\hat{\sigma}_G^2$ significant	Mean	Range	
Aspergillus nidulans	radial growth rate	7	6	66	17-94	a, b
	penicillin titre	7	4	33	0-77	c
	growth yield	2	1	44	22-65	b
Ceratocystis ulmi	radial growth rate	3	3	89	85-92	d, e
	pathogenicity	5	5	47	45-49	d, e
Aspergillus amstelodami	radial growth rate	8	7	47	0-78	b
Neurospora crassa	radial growth rate	7	7	47[c]	30-60[c]	f
Cochliobolus carbonum	radial growth rate	4	4	75	66-86	g
Schizophyllum commune	radial growth rate	6	6	69	40-89	h
Collybia velutipes	radial growth rate	7	7	92	83-96	i

[a] $\hat{\sigma}_G^2$ expressed as a percentage of the total phenotypic variation

[b] References: a, Jinks et al., 1966; b, Caten and Lawrence, unpublished; c, Merrick and Caten, 1975; d, Brasier and Gibbs, 1976; e, Brasier, 1977; f, Papa et al., 1966; g, MacKenzie et al., 1971; h, Simchen, 1966a; i, Croft and Simchen, 1965.

[c] Realised heritabilities

only is adequate, $\hat{\sigma}_G^2$ provides an estimate of the additive genetic variance (D) and $\hat{\sigma}_E^2$ an estimate of the environmental variance (E). In the presence of epistasis, however, $\hat{\sigma}_G^2$ estimates D and I, where I is the non-additive genetic component. D and I can be separately estimated by raising further generations, such as backcrosses or an F_2 (Mather and Jinks, 1971). However, as the F_1 is a segregating generation, raising further generations requires sampling the F_1 and, hence, involves many additional crosses.

Over 50 haploid, bi-parental crosses have been carried out and the progeny assessed for one or more continuous variables (Table II). The majority of crosses involved independent isolates of Ascomycetes, and parental and F_1 generations were included. In the Basidiomycetes *S. commune* and *Collybia velutipes* natural dikaryons in which the parental haploids were already associated were fruited to produce the F_1 (Simchen, 1966a; Croft and Simchen, 1965). Only in *C. velutipes* was it possible to recover the parental haploids from the dikaryons and, hence, to carry out the complete analysis. Significant genetic variation was detected in 89% of the crosses (Table II) and, with a few exceptions, the F_1 progenies showed a continuous, unimodal and frequently near-normal distribution of phenotypes (Jinks *et al.*, 1966; Simchen, 1966a; Merrick and Caten, 1975; Braiser and Gibbs, 1976). The proportion of the total phenotypic variation attributable to genetic effects varied from system to system and from cross to cross within a system (Table II). With radial growth rate where the environment can be carefully controlled and the measurement errors are low, heritability estimates of over 90% have been obtained. In these cases the observed F_1 distribution is a close reflection of the segregation of allelic differences between the parents and, in the absence of heterozygotes, a continuous unimodal F_1 distribution suggests the simultaneous segregation of several allele pairs of similar effect.

Where tests for non-additive variation have been carried out, a simple additive model has proved adequate in most cases (Butcher, 1969; Merrick and Caten, 1975). Non-additive effects have been detected where the parents were of divergent origin (Jinks *et al.*, 1966; Butcher, 1969; Brasier, 1977) and in crosses between independently selected lines (Papa, 1970); in both cases they presumably reflect the breakup of balanced polygenic complexes.

b. *Tetrads*. The hypothesis that continuous variation is determined by the simultaneous segregation at meiosis of several chromosomal genes is central to quantitative genetic theory (Falconer, 1960; Mather and Jinks, 1971). Since individual genes cannot be followed the demonstration of segregation rests upon the relative magnitude of the variances in different generations of defined breeding programmes. In many fungi it is possible to recover all four products of a single meiosis (tetrad analysis) and Pateman and Lee (1960) made use of this property to demonstrate directly the segregation of polygenes affecting ascospore size in *N. crassa*. Considering the eight ascospores within a single ascus there were no significant differences within each pair of mitotic products, but products separated by the first or second division of meiosis differed significantly, directly reflecting the segregation of polygenes during each of the two meiotic divisions. Similarly, Caten, Groth, Person and Dhahi (unpublished) found that first and second division segregation contributed equally to the variation in pathogenic aggressiveness within tetrads of *Ustilago hordei*. Segregation at both meiotic divisions within single tetrads suggests that several genes distributed along the length of the chromosomes are involved. There were significant differences between tetrad means in both *N. crassa* and *U. hordei* indicating interaction among the genes involved, since with purely additive action all tetrad means should be the same (Pateman and Lee, 1960).

2. *Multiple-Mating Programmes*

Estimates of additive and non-additive genetic components of variation in haploids can be derived from multiple-mating programmes. However, since segregation occurs in the F_1 generation, it is necessary to sample the progeny of each cross with the result that the experiments become very large in relation to the additional information provided. For this reason multiple-mating programmes have seldom been used, although Butcher (1969) adopted a 4 x 4 diallel mating programme to demonstrate the importance of non-allelic interactions in crosses between heterokaryon incompatible isolates of *A. nidulans*.

B. *Diploids and Dikaryons*

With diploids and dikaryons each cell contains two copies of the genome and interactions between different alleles at the same locus (*i.e.*, dominance) and between alleles at different loci in the two genomes are potential causes of non-additive genetic effects.

TABLE III. *Penicillin titres of haploid and diploid strains of A. nidulans.*[a]

Haploids		Diploids	
Strain	Titre (u ml^{-1})	Strain	Titre (u ml^{-1})
7-142	7.1 ± 0.3	7-142/B6-27	10.6 ± 0.7
B6-27	14.8 ± 0.7		
7-151	6.8 ± 0.4	7-151/A6-9	7.8 ± 0.6
A6-9	11.7 ± 0.6		

[a] Data from Simpson (1977).

1. Bi-Parental Crosses

In most fungi with stable diploid/dikaryotic phases it is also possible to culture the haploid phase. Where the same character is expressed in both phases of the life cycle, comparison of a diploid/dikaryon with its two haploid components provides a test for net directional non-additive gene effects. Table III gives the penicillin titres of two somatic diploids of *A. nidulans* and of their respective haploid components. Additive gene action will account for diploid 7-142/B6-27 whose titre did not differ from the mean of its haploid components. The titre of diploid 7-151/A6-9, however, differed from the mean of its haploid components, but not from that of its low component (7-151), suggesting directional dominance for low titre (Simpson, 1977). Very large non-additive effects have been revealed by comparisons of the radial growth rates of dikaryons of *S. commune* and *C. velutipes* with the mean growth rates of their component monokaryons (Simchen and Jinks, 1964; Simchen, 1965, 1966b). These were larger than the non-additive component among populations of dikaryons synthesized in multiple-mating programmes (see section II, B, 2) suggesting that different gene systems determine growth rate in the monokaryotic and dikaryotic phases (Simchen and Jinks, 1964; Simchen, 1965).

2. Multiple-Mating Programmes

The stable diploid/dikaryotic association in fungi is equivalent to the normal situation in higher organisms, where observations are made exclusively on the diploid phase, and lends itself to the same multiple-mating programmes and the accompanying analyses. However, since the haploid phase can be separately cultured, a single "gamete" can be cloned and used for an indefinite number of crosses, thereby bypassing the need to develop inbred lines. Many Basidiomycetes possess self-incompatibility systems (Raper, 1966) which prevent the adoption of diallel programmes. However, a design initially developed for plant breeding programmes (North Carolina Experiment II; Comstock and Robinson, 1952) can be adopted and provides considerable information (Simchen and Jinks, 1964). A sample of m monokaryons of one mating type ("mothers") is crossed in all combinations with a sample of f monokaryons of a compatible mating type ("fathers") and the resulting $m\,f\,F_1$ dikaryons are cloned and compared in a suitably replicated experiment. A two-way analysis of variance tests for and provides estimates of additive and non-additive genetic components (Table IV). Where the monokaryotic parents are derived from a single dikaryon and, hence, within sampling error, each segregating locus is represented by two equally frequent alleles, the sufficiency of a model involving only additive, dominance, and environmental variation can be tested by examining the relationship between the variances and covariances of members of half-sib families (Simchen and Jinks, 1964).

The conclusions from the application of this crossing programme to the investigation of dikaryotic characters in three Basidiomycete species are summarized in Table V. Several generalizations can be drawn. Firstly, significant additive (σ_P^2) and non-additive (σ_I^2) effects have been detected in all populations examined, with the exception of a known inbred dikaryon of *U. hordei*. Secondly, with one exception, the additive genetic component is larger than the non-additive component. Thirdly, in *S. commune* and *C. velutipes* non-additive effects are more important for the fruiting characters than for radial growth rate. Fourthly, considering natural dikaryons of *S. commune*, the non-additive variation is due to dominance only for growth rate, but epistasis is also involved for the fruiting characters. Fifthly, the genetic architecture of growth rate in populations of *S. commune* derived from unrelated monokaryons (synthetic populations) is more complex than that in populations obtained from a single natural dikaryon; this is consistent with the independent evolution of the unrelated genotypes (Simchen, 1967).

TABLE IV. *Analysis of variance for F_1 diploids/dikaryons produced by a multiple-mating programme.*[a]

Source of variation	Degrees of freedom	Expected mean square
Between "fathers"	$f - 1$	$\sigma_E^2 + n\sigma_I^2 + n\,m\sigma_{P1}^2$
Between "mothers"	$m - 1$	$\sigma_E^2 + n\sigma_I^2 + n\,f\sigma_{P2}^2$
"fathers" × "mothers"	$(f - 1)(m - 1)$	$\sigma_E^2 + n\sigma_I^2$
Replicates	$f\,m\,(n - 1)$	σ_E^2

[a] f = number of "fathers"; m = number of "mothers" n = number of replicates; σ_E^2 = environmental component of variation; σ_I^2 = non-additive genetic component; σ_{P1}^2 and σ_{P2}^2 = additive genetic components due to "fathers" and "mothers" respectively.

III. SELECTION

A. *Response*

Fungi possess two major advantages for selection experiments: (1) strains can be stored making it possible to compare genotypes from all generations in a single environment at the end of the selection programme, (2) since individual genotypes can be cloned the genetic and environmental components of variation can be estimated and their changes from generation to generation followed. Selection for continuously varying characters has been carried out on both the haploid (Simchen, 1966a; Papa *et al.*, 1966; Merrick, 1975a) and diploid/dikaryotic phases (Pateman, 1959; Lee, 1962; Connolly and Simchen, 1973). In all cases a significant response was obtained, although the rate, duration, and final magnitude of response varied from study to study, depending upon the organism, character, breeding system, and selection intensity involved. Considered together these studies with fungi have revealed most of the features of selection previously encountered in higher organisms, *e.g.* asymmetrical responses, loss of fertility, plateaus, correlated responses (Mather and Harrison, 1949; Lerner, 1958; Falconer, 1960).

Selection on haploids is simplified by the absence of dominance and, other factors being favourable, should produce a rapid symmetrical response with accompanying decline in the genetic variance (Figure 1). This idealized response has not

TABLE V. Genetic variation for dikaryotic characters in Basidiomycetes.

Species and Character	Origin	No. experiments	% Phenotypic variation[a] σ_P^2	σ_I^2	σ_E^2	Exper. with Model[b] Sufficient	Not Sufficient	Reference[c]
S. commune								
growth rate	natural dikaryons	7	51	18	31	7	0	a, b
fruiting time	synthetic populations	8	61	17	22	4	4	c
fruiting time	natural dikaryons	6	54	26	20	1	5	a
fruit weight	natural dikaryons	3	38	36	26	1	2	a
C. velutipes								
growth rate	synthetic population	1	71	21	8	1	0	d
no. primordia	synthetic population	1	51	38	11	1	0	d
no. sporophores	synthetic population	1	30	37	33	0	1	d
fruiting time	synthetic population	1	48	31	21	1	0	d
U. hordei								
pathogenicity	synthetic population	1	42	9	49	--	--	e
	2 teliospores	1	44[d]	21[d]	35	--	--	f
	inbred dikaryon	1	0[d]	8[d]	92	--	--	g
	synthetic dikaryon	1	29	20	51	1	0	g

[a] Mean of all experiments. See Table IV for definitions; $\sigma_P^2 = \sigma_{P1}^2 + \sigma_{P2}^2$
[b] Model involves additive, dominance, and environmental effects only.
[c] References: a, Simchen, 1966b; b, Williams et al., 1976; c, Simchen, 1967; d, Simchen, 1965; e, Emara, 1972; f, Emara and Sidhu, 1974; g, Caten et al., unpublished.
[d] Not significantly different from zero; "--", not tested.

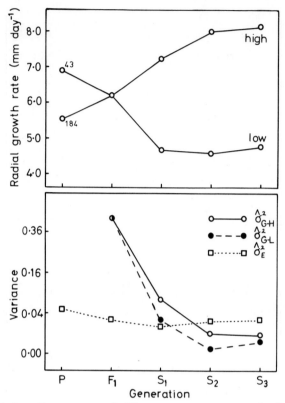

FIGURE 1. Response of the progeny mean and the genetic ($\hat{\sigma}_G^2$) and environmental ($\hat{\sigma}_E^2$) variance components to selection for radial growth rate among the progeny of Aspergillus nidulans cross 43 × 184. (Adapted from Caten and Jinks, 1976).

always been obtained. Monokaryotic lines of *S. commune* selected for low growth rate showed a greater response and maintained higher genetic variances than the high lines (Simchen, 1966a). This asymmetry was attributed to the instability of the low growth rate genotypes and to linkage of genes affecting growth rate to the mating-type factors (Simchen, 1966a; Connolly and Simchen, 1968). A similar asymmetrical response was obtained by Connolly and Simchen (1973) when selecting for dikaryotic growth rate in *S. commune*. In this case the higher response for low growth was attributed to the combined effects of a greater selection differential and inbreeding depression.

B. *Selection Limits*

Plateaus in selection experiments arise from a variety of causes (Falconer, 1960) which may be broadly divided into physiological and genetic limits. These origins are clearly illustrated by comparison of the results of selection for radial growth rate in *N. crassa* and for penicillin titre in *A. nidulans*. Both studies involved the haploid phase and selection was for high expression only in a number of independent lines. Response was rapid at first but soon plateaued, the observed falls in the genetic variance suggesting that a genetic limit had been reached (Papa *et al.*, 1966; Merrick, 1975a). In both cases crosses between independent selection lines released new genetic variation; however, while this permitted a further increase in penicillin titre (Merrick, 1975b) renewed selection was not effective for growth rate (Papa, 1970). Furthermore, in the *N. crassa* growth rate experiments reverse selection revealed the persistence of genetic variation in high selection lines long after they had stopped responding (Papa *et al.*, 1967). Thus, the plateaus in the original selection lines appeared due to genetic limits for penicillin titre in *A. nidulans*, but to physiological limits for radial growth rate in *N. crassa*.

C. *Correlated Responses*

Selection in higher organisms is frequently accompanied by reductions in fertility and by developmental instability (Mather and Harrision, 1949; Lerner, 1954; Thoday, 1958). Changes in both these characteristics have been encountered in selection experiments with fungi. Infertility was particularly evident in the experiments on *N. crassa* where it was manifested by the formation of fewer perithecia and failure of ascospore maturation and germination (Pateman, 1959; Lee and Pateman, 1961; Papa *et al.*, 1966). In some lines fertility could be restored by relaxing selection for a few generations (Papa *et al.*, 1966). Increased developmental instability was a component of the response to selection for ascospore size in *N. crassa* (Pateman, 1959; Lee, 1962) and contributed to the maintenance of genetic variance in the low growth rate selections of *S. commune* (Simchen, 1966a; Connolly and Simchen, 1973). Since many of the experiments involved the haploid phase, infertility and developmental instability must be a property of the selected genotypes and not of homozygosity *per se*.

IV. THE NATURE OF POLYGENIC SYSTEMS

A. *The Number of Effective Factors*

The observation of a continuous distribution of phenotypes in a segregating population is frequently interpreted to indicate that a large number of genes, or more strictly effective factors, are involved. However, this conclusion is not justified on these grounds alone, since in diploids two or three loci are sufficient to produce distributions which are in practice indistinguishable from normal distributions (Thoday and Thompson, 1976). With haploid fungi this possibility of confusing relatively simple genetic control with a polygenic system is reduced by both the absence of intermediate heterozygotes and the ability to minimise the environmental variance through the use of controlled environments and clonal replication. It is, therefore, far more likely that in these systems the segregation of one or two pairs of alleles with an effect markedly greater than any others will be revealed by a detectably multimodal distribution. This section is concerned with information on the nature of polygenic systems obtained from studies with haploid fungi.

Continuous, unimodal distributions were observed in the F_1 of all except four of the fifty crosses which gave a significant genetic component (Table II). The four exceptional crosses showed bimodal distributions, indicating the segregation of a single factor of major effect which accounted for a large part, but not all, of the genetic variation. In general, therefore, the continuous variation in natural fungal populations appears to be determined by several genes of similar effect and allelic differences with a major effect occur only at a low frequency. Confirmation of the involvement of several genes requires the isolation of separable loci (Breese and Mather, 1957; Thoday, 1961) or the demonstration of continued segregation through successive generations of inbreeding (Jinks and Towey, 1976), techniques which have not yet been applied to fungi. Nevertheless, several indirect considerations suggest that more than a few loci are involved for each character in each population.

An estimate of the number of effective factors (k) segregating in a haploid cross is provided by comparison of the difference between the extreme F_1 progeny individuals ($P_H - P_L$) and the genetic variance (σ_G^2) as

$$k = \frac{(P_H - P_L)^2}{4\sigma_G^2}$$

(Chovnick and Fox, 1953; Croft and Simchen, 1965; Mather and Jinks, 1971). Since with many genes the extreme genotypes will be rare, the progeny sample size limits this estimate. With samples of the size generally used (75-100) the method can only discriminate less than five factors. With the exception of one cross in *Collybia velutipes* which was segregating for a major gene (Croft and Simchen, 1965), the estimates have either equalled or exceeded this statistical limit, indicating that each cross was segregating for five or more factors influencing the character in question (Caten and Jinks, 1976). The involvement of several genes in F_1 progenies is borne out by the cumulative response observed in selection experiments initiated from a single cross (Simchen, 1966a; Papa et al., 1966; Merrick, 1975a). Considering a population of independent isolates the number of polymorphic loci affecting the character may be much greater than that segregating in any one cross, since different loci may be involved in different crosses. That this is the case is suggested by crosses between lines selected for similar phenotype from different initial bi-parental crosses. These released as much genetic variation as crosses between independent, unselected isolates indicating that different alleles had been fixed in each line (Papa, 1970; Merrick, 1975b). It seems that each of the initial crosses was segregating for different genes and/or alleles affecting the character in question and, therefore, that the variation in the total population is determined by a large number of genes.

The analysis of crosses can at best only reveal how many genes influencing the character are polymorphic in the population; it provides little information on the number potentially capable of exerting an effect. Evidence on this point may be drawn from mutational studies, however, since these permit examinations of a much wider range of alleles than survive natural selection. Macdonald et al. (1963) found that 60% of mutations to auxotrophy in *Penicillium chrysogenum* affected penicillin titre, even when their growth requirement was supplemented. Assays of random survivors of mutagenic treatment in *A. nidulans* indicate that more than 5% have penicillin titres significantly different from their parent strain (Simpson and Caten, 1979). Comparisons of this frequency with that of mutants produced by the same treatments in test systems involving known numbers of loci provide estimates of the number of loci that can influence penicillin titre in this species. These estimates range from 75 to 2250 with a mean of 982, depending upon the particular mutagenic treatment used (Simpson and Caten, 1979). While many assumptions are involved, these estimates emphasise that a complex character, such as penicillin titre, is potentially sensitive to allelic substitutions at many genes. Considering this, together with the high frequency of enzymic polymorphism revealed by electrophoretic surveys of natural populations (Selander, 1976), most independent isolates would be expected to differ at several genes capable of influencing any complex variable. The continuous progeny distributions obtained in crosses are clearly consistent with this expectation.

B. *The Isolation of Individual Effective Factors*

1. Location of Effective Factors. The above considerations (section IV, A) concerning the number of effective factors say nothing about the relative magnitude of their effects. For this purpose it is necessary to isolate individual factors. This is facilitated by chromosome assay techniques whereby the net effect of all the genes within a chromosome is assessed (Mather, 1942; Mather and Harrison, 1949) and by the analysis of intrachromosomal recombinants (Breese and Mather, 1957; Thoday, 1961). The parasexual cycle (Roper, 1966) offers a system of chromosome assay in fungi. As yet, however, exploitation of this system has been prevented by the vegetative (heterokaryon) incompatibility of unrelated fungal isolates (Caten and Jinks, 1966) which blocks the necessary somatic diploid formation. As a result the effective factors isolated in fungi have been discovered fortuitously.

2. Factors of Major Effect. Those crosses which revealed the segregation of an allele pair of major effect provide an opportunity for locating a component of a polygenic system. Isolates of A. *nidulans* belonging to heterokaryon-compatibility group F (h-c F) produce no detectable penicillin activity (Holt and Macdonald, 1968). The F_1 of a cross between the h-c F isolate 136 and isolate 109 (14 u penicillin ml^{-1}) segregated into two classes in a 1:1 ratio (Figure 2). This bimodal distribution contrasts markedly with that shown by the same character in other crosses (Merrick and Caten, 1975) and indicates the presence of a mutant allele determining nonproductivity in isolate 136, and probably in all h-c F isolates. Cole *et al.* (1976) showed that this mutation is allelic with laboratory-induced mutants for penicillin non-production mapping at the *npeA* locus on chromosome VI. The low productivity (<1 u ml^{-1}) of h-c G isolates is also due to a natural mutant allele at the *npeA* locus, but this allele is different from that in h-c F isolates (Cole *et al.*, 1976). Thus, at least three alleles at the *npeA* locus exist in the A. *nidulans* population, one permitting high titres, one limiting titres to a low level, and one blocking penicillin

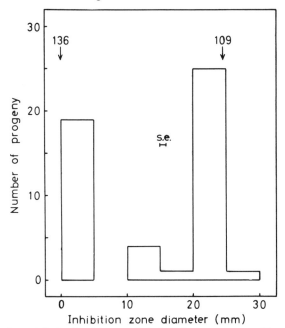

FIGURE 2. *Distribution of inhibition zone diameter (penicillin bioassay) among 50 F_1 progeny of Aspergillus nidulans cross 109 × 136.*

production altogether. Therefore, depending upon the particular strains crossed, this locus may have no effect, a minor effect, or a major effect on penicillin titre.

3. *Linkage to Markers.* In addition to the continuous variable many haploid crosses were segregating for one or more natural or induced markers. Provided these have no pleiotropic effect on the variable, comparison of the means of the different marker genotypes in the F_1 provides a test for linkage of effective factors to the markers. Detection of linkage in this way will depend upon the contribution of the linked component relative to that of the rest of the genome and upon the degree of linkage. Generally differences between marker genotypes have not been found, confirming the absence of both pleiotropy and linkage (Jinks *et al.*, 1966; Simchen, 1966a). A surprising number of crosses, however, have revealed effective factors linked to markers in the F_1 (Lee and Pateman, 1959; Simchen, 1966a; Connolly and Simchen, 1968; Caten and Jinks, 1976). Linkage of growth rate genes to the *aurescent* locus (*aur*) on chromosome I of *N. crassa* could not be detected in the F_1 but was apparent when *aur* F_1 strains were backcrossed to the aur^+ parent, thereby reducing the background genetic

variance (Papa, 1971).

4. *Magnifying the Effect of Individual Allelic Substitutions.* The isolation of factors in a polygenic system would be greatly facilitated by a method which selectively magnified the effect of an individual factor. On the assumption that missense mutations leading to enzymic polymorphisms are a major component of quantitative genetic variation, it might be possible to increase the effect of particular allelic substitutions by growing segregating populations at high temperatures, since laboratory-induced missense mutations frequently lead to the production of temperature sensitive proteins

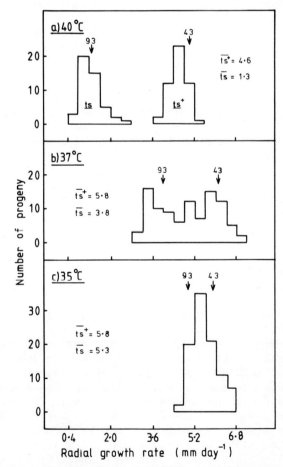

FIGURE 3. *Distribution of radial growth rate among 96 F_1 progeny of Aspergillus nidulans cross 43 × 93 grown at three temperatures.*

(Hayes, 1968). Accordingly, natural isolates of *A. nidulans* were screened for temperature sensitivity and one (isolate 93) which grew well at 30°C but poorly at 40°C was identified. Isolate 93 was crossed to isolate 43, which grew well at both temperatures, and the growth rates of the same 96 F_1 progeny were determined at each of a range of temperatures. At 40°C the F_1 progeny fell into two, equally frequent, discrete classes indicating the segregation of a major gene, designated *ts* (Figure 3a). This gene still had a major effect at 37°C resulting in an F_1 distribution which, although continuous, was bimodal (Figure 3b). At 35°C and below, however, the familiar normal distribution with a highly significant genetic component was obtained and the *ts* gene had only a minor effect (Figure 3c). Thus, depending upon the environment, the same pair of alleles may act as a minor component of a polygenic system, as a factor of major effect in a polygenic system, or as a major gene.

V. PHYSIOLOGICAL COMPONENTS OF QUANTITATIVE VARIATION

In the course of the work at Birmingham on quantitative genetic variation in *A. nidulans* attempts have been made to identify major physiological components of the characters studied. These have not met with great success. Rather, as the following two examples show, they have emphasised the complexity and frequently unexpected nature of the component variables.

A. *Radial Growth Rate in Aspergillus nidulans*

The differences in radial growth rate among independent isolates and between contrasting selection lines were far greater on a simple synthetic medium (minimal) than on a rich medium containing organic supplements (complete). The slow growing strains may, therefore, be compared to leaky auxotrophic mutants, suggesting that their poor performance might originate from a biosynthetic deficiency which was by-passed in the supplemented medium. Experiments were undertaken to identify this postulated deficiency by fractionating the complete supplement. These revealed the complex organic constituents yeast extract and peptone as the active fractions and defined amino acids, vitamins, purines and pyrimidines were largely ineffective. Thus, slow radial growth rate could not be related to any simple nutritional deficiency.

Colony radial growth rate is a function of specific growth

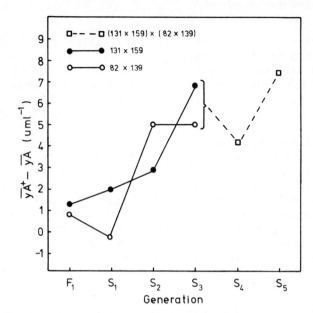

FIGURE 4. *Changes in the effect of the yA allele on penicillin titre during selection for increased titre in Aspergillus nidulans.*

rate (rate of increase in dry weight) and the width of the colony's peripheral growth zone (Trinci, 1971). The specific growth rates of fast and slow radial growth rate selections were the same, indicating that differences in the peripheral growth zone were responsible for the natural variation in radial growth rate. A number of developmental processes affecting the peripheral growth zone have been identified, including hyphal branching pattern, frequency of septum formation and rate of septal plugging (Trinci, 1971; Trinci and Collinge, 1973). However, the physiological and biochemical control of these processes is not presently understood and they offer little hope for a fundamental explanation of a continuous variable.

B. *Penicillin Titre in Aspergillus nidulans*

The majority of crosses carried out during the investigation of the inheritance of penicillin titre in *A. nidulans* were segregating for the yellow conidial colour marker, *yA*. The initial crosses indicated that *yA* produced only a small reduction in titre (Merrick and Caten, 1975). As selection progressed and the mean titre increased, however, there was an increasing divergence between the mean titres of green and yellow spored offspring (Merrick, 1975a, 1975b; Figure 4). The mean titre of yA segregants was always lower than that of yA^+ segregants irrespective of the parent introducing the yA allele, indicating that this was a pleiotropic effect (Merrick, 1975a). It was not immediately apparent how yA mutations, which result from a deficiency in a p-diphenol oxidase (Clutterbuck, 1972), might reduce penicillin titre. However, it was noted that yA strains produced less dense conidial inocula for the shake flask fermentations than did yA^+ strains. Standardization of the inocula removed the difference in titre confirming an indirect effect of yA through inoculum density (Merrick, 1973). This result was surprising since experiments with unselected isolates had indicated that inoculum density and penicillin titre were independent over the range involved. Repeat of these experiments using both unselected isolates and high-titre selections showed that the latter were considerably more sensitive to inoculum density (Merrick, 1973). Given these different responses to inoculum density it can be seen how a mutation reducing conidiation would have a large effect in a high-titre background, but little effect in a low one.

Although the $yA:yA^+$ allelic difference was introduced artificially it was one factor in the polygenic system influencing penicillin titre in these crosses. As such it offers a model of the way in which genes with primary effects far removed from the immediate physiology of the variable may exert a pronounced effect. Furthermore, these indirect effects can interact markedly with other factors in the genotype.

VI. CONCLUSIONS

A variety of biometrical techniques are available for the genetic analysis of quantitative variation in fungi. The techniques applicable in any particular instance are determined by the biology of the organism, especially the ploidy and mating system, and by the nature of the information required. Use of these techniques has provided valuable information on the inheritance of fungal characters of fundamental and applied interest, but to date has added little to our understanding of quantitative genetics. However, because of their many technical advantages, the potential contribution of fungi to this area of genetics is great, particularly with regard to the nature and physiology of polygenic systems. Routine methods for the isolation and mapping of individual effective factors remain to be developed and exploited. Despite this deficiency the information already obtained provides some insight into the nature of polygenic systems.

The high rate of induction of continuous variation and the frequent pleiotropy of major markers suggest that many variables may be influenced by a significant part of the genome. The variation in complex characters in natural populations also appears to be determined by many genes, as evidenced by statistical estimates, selection experiments, tetrad analysis, and the frequency of detection of linkage to markers. However, the number of allelic differences and the magnitude of their effects may clearly vary from cross to cross and population to population. Discrete or continuous variation are not properties of particular characters but depend upon the complexity of the underlying genetic differences and the degree of environmental control, in each case. For the same character, in some crosses most of the variation results from the segregation of one or a few allelic pairs with major effects, while in others a large number of factors of individually small effect are involved. In a natural population, therefore, alleles with a range of magnitudes of effect are to be expected, and, depending upon the pair of alleles involved, the same locus may have a major or a minor effect. Furthermore, a particular pair of alleles may have a major effect in one environment but a minor effect in another, or may exert a major effect on one character but a minor effect on another. Thus, it seems that the same loci may be involved in discrete and continuous variation and that the individual factors in a polygenic system are allelic variants with only a minor effect on the activity of the gene product. In these cases the distinction between major genes and polygenes is operational rather than constitutional and is determined by the nature of the experimental material and the methods of investigation. However, while isoallelic variation of structural genes is probably a major source of quantitative genetic variation, it is not the sole component of polygenic systems. Heterochromatic chromosome segments which lack demonstrable major genes have polygenic activity in higher organisms (Mather, 1944) and control genes may also be involved (Thoday, 1977).

As products of whole organisms with integrated developmental, physiological and biochemical pathways, phenotypic characters removed from the primary gene products are likely to show continuous variation in all but the simplest situations. This variation may arise at many points, some close to the immediate physiology and biochemistry of the character itself and others far removed from this area.

ACKNOWLEDGMENTS

I am deeply indebted to present and former colleagues in the Department of Genetics, University of Birmingham for many valuable discussions on quantitative genetics both in fungi and higher organisms. I thank Professor J.L. Jinks, Professor K. Mather, and Dr. J. Croft for their helpful comments on the manuscript and Jean Lawrence and Hazel Howell for their excellent technical assistance. The work on growth rate and penicillin production in *A. nidulans* was supported by a grant from the British Science Research Council.

REFERENCES

Alikhanian, S.I. (1970). *Curr. Top. Micro. Immunol.* 53, 91-148.
Brasier, C.M. (1970). *Amer. Nat.* 104, 191-204.
Brasier, C.M. (1977). *Trans. Brit. Mycol. Soc.* 68, 45-52.
Brasier, C.M., and Gibbs, J.N. (1976). *Ann. Appl. Biol.* 83, 31-37.
Breese, E.L., and Mather, K. (1957). *Heredity* 11, 373-395.
Burnett, J.H. (1975). "Mycogenetics." Wiley, London.
Butcher, A.C. (1969). *Heredity* 24, 621-631.
Caten, C.E., and Day, A.W. (1977). *Ann. Rev. Phytopathol.* 15, 295-318.
Caten, C.E., and Jinks, J.L. (1966). *Trans. Brit. Mycol. Soc.* 49, 81-93.
Caten, C.E., and Jinks, J.L. (1976). *In* "Second International Symposium on the Genetics of Industrial Microorganisms" (K.D. Macdonald, ed.), pp. 93-111. Academic Press, London.
Chovnick, A., and Fox, A.S. (1953). *Amer. Nat.* 87, 263-267.
Cole, D.S., Holt, G., and Macdonald, K.D. (1976). *J. Gen. Microbiol.* 96, 423-426.
Comstock, R.E., and Robinson, H.F. (1952). *In* "Heterosis" (J.W. Gowen, ed.), pp. 494-516. Iowa State College Press, Ames.
Connolly, V., and Simchen, G. (1968). *Heredity* 23, 387-402.
Connolly, V., and Simchen, G. (1973). *Genet. Res.* 22, 25-36.
Croft, J.H., and Jinks, J.L. (1977). *In* "Genetics and Physiology of *Aspergillus*" (J.E. Smith and J.A. Pateman, eds.) pp. 339-360. Academic Press, London.
Croft, J.H., and Simchen, G. (1965). *Amer. Nat.* 99, 451-462.
Clutterbuck, A.J. (1972). *J. Gen. Microbiol.* 70, 423-435.
Day, P.R. (1974). "Genetics of Host-Parasite Interaction." Freeman, San Francisco.

Elander, R.P., and Espenshade, M.A. (1976). *In* "Industrial Microbiology" (B.M. Miller and W. Litsky, eds.), pp. 192-256. McGraw-Hill, New York.
Emara, Y.A. (1972). *Can. J. Genet. Cytol. 14*, 919-924.
Emara, Y.A., and Sidhu, G. (1974). *Heredity 32*, 219-224.
Esser, K., and Kuenen, R. (1967). "Genetics of Fungi." Springer Verlag, New York.
Falconer, D.S. (1960). "Quantitative Genetics." Oliver and Boyd, Edinburgh.
Fincham, J.R.S., and Day, P.R. (1971). "Fungal Genetics" (3rd edition). Blackwell, Oxford.
Hayes, W. (1968). "The Genetics of Bacteria and Their Viruses" (2nd edition). Blackwell, Oxford.
Holt, G, and Macdonald, K.D. (1968). *Antonie van Leeuwenhoek 34*, 409-416.
Jinks, J.L., and Towey, P. (1976). *Heredity 37*, 69-81.
Jinks, J.L., Caten, C.E., Simchen, G., and Croft, J.H. (1966). *Heredity 21*, 227-239.
Lee, B.T.O. (1962). *Aust. J. Biol. Sci. 15*, 160-165.
Lee, B.T.O., and Pateman, J.A. (1959). *Nature (London) 183*, 698-699.
Lee, B.T.O., and Pateman, J.A. (1961). *Aust. J. Biol. Sci. 14*, 223-230.
Lerner, I.M. (1954). "Genetic Homeostasis." Oliver and Boyd, Edinburgh.
Lerner, I.M. (1958). "The Genetic Basis of Selection." Wiley, New York.
Macdonald, K.D., Hutchinson, J.M., and Gillett, W.A. (1963). *J. Gen. Microbiol. 33*, 365-374.
MacKenzie, D.R., Nelson, R.R., and Cole, H. (1971). *Phytopathology 61*, 471-475.
Mather, K. (1942). *J. Genet. 43*, 309-336.
Mather, K. (1944). *Proc. Roy. Soc. London, B, 132*, 308-332.
Mather, K., and Harrison, B.J. (1949). *Heredity 3*, 1-52 and 131-162.
Mather, K., and Jinks, J.L. (1971). "Biometrical Genetics" (2nd edition). Chapman and Hall, London.
Merrick, M.J. (1973). "Quantitative Genetics of Antibiotic Production in *Aspergillus nidulans*." Ph.D. thesis, University of Birmingham, England.
Merrick, M.J. (1975a). *J. Gen. Microbiol. 91*, 278-286.
Merrick, M.J. (1975b). *J. Gen. Microbiol. 91*, 287-294.
Merrick, M.J., and Caten, C.E. (1975). *J. Gen. Microbiol. 86*, 283-293.
Papa, K.E. (1970). *Can. J. Genet. Cytol. 12*, 1-9.
Papa, K.E. (1971). *Genetica 42*, 181-186.
Papa, K.E., Srb, A.M., and Federer, W.T. (1966). *Heredity 21*, 595-613.

Papa, K.E., Srb, A.M., and Federer, W.T. (1967). *Heredity* 22, 285-296.
Pateman, J.A. (1959). *Heredity 13*, 1-21.
Pateman, J.A., and Lee, B.T.O. (1960). *Heredity 15*, 351-361.
Raper, J.R. (1966). *In* "The Fungi" (G.C. Ainsworth and A.S. Sussman, eds.), Vol. II, pp. 473-511. Academic Press, New York.
Roper, J.A. (1966). *In* "The Fungi" (G.C. Ainsworth and A.S. Sussman, eds.), Vol. II, pp. 589-617. Academic Press, New York.
Schwarzbach, E., and Wolfe, M.S. (1976). *In* "Barley Genetics III" (G. Gaul, ed.), pp. 426-432. Karl Thiemig, Munich.
Selander, R.K. (1976). *In* "Molecular Evolution" (F.J. Ayala, ed.), pp. 21-45. Sinauer, Sunderland, Mass.
Simchen, G. (1965). *Genetics 51*, 709-721.
Simchen, G. (1966a). *Heredity 21*, 241-263.
Simchen, G. (1966b). *Genetics 53*, 1151-1165.
Simchen, G. (1967). *Evolution 21*, 310-315.
Simchen, G., and Jinks, J.L. (1964). *Heredity 19*, 629-649.
Simpson, I.N. (1977). "A Quantitative Investigation of the Genetics of Penicillin Production in Mutation Selected Lines of *Aspergillus nidulans*." Ph.D. thesis, University of Birmingham, England.
Simpson, I.N., and Caten, C.E. (1979). *J. Gen. Microbiol. 110*, 1-12.
Thoday, J.M. (1958). *Heredity 12*, 401-415.
Thoday, J.M. (1961). *Nature (London) 191*, 368-370.
Thoday, J.M. (1977). *Heredity 39*, 427-428.
Thoday, J.M., and Thompson, J.N., Jr. (1976). *Genetica 46*, 335-344.
Trinci, A.P.J. (1971). *J. Gen. Microbiol. 67*, 325-344.
Trinci, A.P.J., and Collinge, A.J. (1973). *Arch. Mikrobiol. 91*, 355-364.
Williams, S., Verma, M.M., Jinks, J.L., and Brasier, C.M. (1976). *Heredity 37*, 365-375.

POLYGENIC VARIATION IN NATURAL POPULATIONS
OF *DROSOPHILA*

P. A. *Parsons*

Department of Genetics and Human Variation
La Trobe University
Bundoora, Victoria, Australia

I. INTRODUCTION

The aim of this chapter is to discuss the range of variation, within and between species, in natural populations of *Drosophila* for traits involving differential responses to the environment, mainly extremes of temperature, desiccation, and alcohol concentrations. Traits involved in population continuity are under discussion, rather than the morphological quantitative traits studied in many of the classic experiments of quantitative genetics. Early work on morphological traits employed either the biometrical approach which began with Fisher (1918), or rather less commonly (see Payne, 1918) the biologically informative non-biometrical analysis of quantitative variables leading in recent years to the actual location of polygenic activity (Mather and Harrison, 1949; Thoday, 1961, 1967). A comparative review of the two approaches appears in Lee and Parsons (1968). This latter article also provides a preliminary discussion of the assessment and exploitation of variability within and among isofemale strains set up from single inseminated females from natural populations using scutellar chaeta numbers in *D. melanogaster* as an example. Combined with both the biometrical and non-biometrical approaches, it will be shown that isofemale strains provide a powerful technique for the assessment of variation in natural populations.

For many traits, but in most detail for scutellar chaeta number in *D. melanogaster*, it has been shown that isofemale strains maintain consistent differences for many generations after being set up in the laboratory; these differences clearly arise from genetic differences among the founder females. Hence, this forms a way of sampling variation in populations. This variation can then be studied genetically by the biometrical technique of diallel crossing among extreme strains, or by locating polygenic activity in extreme strains by non-biometrical methods. In addition, directional selection based upon extreme isofemale strains and hybrids between them can be remarkably effective in providing bigger differences for study than are usual for experiments based upon unsampled natural populations (Parsons, 1975). Many quantitative traits have been investigated using isofemale strains. These include: *morphological* -- scutellar, sternopleural, and abdominal chaeta number; egg and wing length; mean wet and dry body weights; *behavioral* -- mating speed, duration of copulation, levels of sexual isolation, dispersal activity towards light, phototaxis, maze-learning ability, and reaction to alcohol; *environmental stresses* -- resistance to irradiation with ^{60}Co-γ rays, ether, chloroform, CO_2, and phenylthiourea, as well as traits of ecological significance including high and low temperatures, desiccation, and ethyl alcohol (see Parsons, 1977a).

Given that the aim is to survey variability with natural populations in mind, traits that directly relate to environmental variables in nature are of particular interest. It is reasonable, therefore, to commence with the climate-related variables of temperature and desiccation extremes.

II. RESISTANCE TO CLIMATE-RELATED ENVIRONMENTAL EXTREMES

The success of a *Drosophila* population depends basically upon its adaptation to annual climatic cycles (Dobzhansky, 1950) -- a view favored by Andrewartha and Birch (1954) when considering insect populations in general, especially those in temperate regions. The annual cycle of the temperate zone provides two major density-independent climatic stresses, namely, (1) a combination of high temperature and desiccation stress, and (2) low temperatures, both of which may have major effects upon population sizes (see Weins, 1977). For both of these stress categories it is important to distinguish between conditions for resource utilization (feeding, breeding) and survival, although they would be expected to be correlated (Parsons, 1978a).

Most relevant data concern survival (or mortality) rates. For example, high temperature-sensitive strains of *D. melanogaster* set up from the wild are known, and isofemale strain heterogeneity has been described for ability to withstand desiccation and high temperature shocks (Parsons, 1973). Populations of both *D. melanogaster* and *D. simulans* from Uganda, an area of high temperatures, are more resistant to high temperatures than are population samples from less extreme environments (Tantawy and Mallah, 1961). Temperature "races" occur in *D. funebris* from northern Europe, northern Africa and Asiatic Russia, such that resistance to high and low temperatures corresponds to climate (Dubinin and Tiniakov, 1947). These cosmopolitan species, therefore, provide evidence of being subdivided into "races" with respect to extreme environments.

Mortality after exposure to the cold stress of $-1°C$ has been measured for isofemale strains of *D. melanogaster* and *D. simulans* from Townsville, north Queensland (a subtropical climate with a hot damp summer and warm dry winter, at latitude $20°S$, Figure 1) and Melbourne (a temperate climate with a hot, relatively dry summer and cool damp winter, at latitude $38°S$, Figure 1). There was significant variation among isofemale strains within populations as expected (Parsons, 1977b). Although there was some overlap among populations, the sequence of mean mortalities (Table I) was *melanogaster* (Townsville) > *simulans* (Townsville) > *simulans* (Melbourne) >> *melanogaster* (Melbourne). In particular the *simulans* populations did not differ significantly, while the *melanogaster* populations did ($P < 0.001$). Utilizing isofemale strains enables some assessment of intrapopulation variability derived directly from the wild which can be assessed in relation to interpopulation differences. In Melbourne, resource utilization effectively ceases during winter months, since activity, feeding, and mating of these species effectively ceases at about $12°C$ and below (McKenzie, 1975a), whereas there is no such restriction in the Townsville population as can be shown from the meteorological data in Figure 1.

The clear advantage of *D. melanogaster* in Melbourne over *D. simulans* can almost certainly be correlated with the

TABLE I. *Mean percentage mortalities of D. melanogaster and D. simulans after 48 hours exposure to $-1°C$ (after Parsons, 1977b).*

	Melbourne		Townsville	
	male	female	male	female
D. melanogaster	33.3	25.4	78.4	82.5
D. simulans	95.8	93.4	89.6	78.0

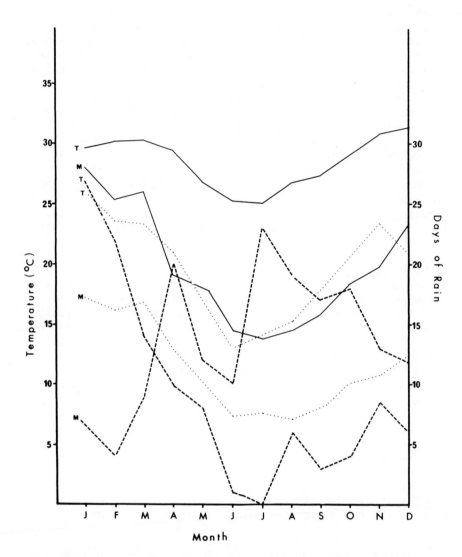

FIGURE 1. Mean monthly maximum and minimum temperatures and mean number of days of rain for Melbourne (M) and Townsville (T), based on meteorological records published by the Australian Government Publications Service for 1974. Maximum temperature (———), mean minimum temperature (•••••), days of rain (-----).

earlier build-up of *D. melanogaster* in Spring compared with *D. simulans* (McKenzie and Parsons, 1974a), leading to a maximum population of *D. melanogaster* in late Spring and of *D. simulans* in late summer. While the genetic basis of cold-resistance was not studied in these populations, this stress evidently involves differing underlying physiological (and genetic) processes from desiccation stress, since the correlation coefficient among isofemale strains for mortalities to the two stresses is close to zero. Indeed, levels of resistance to desiccation stress are associated with body weight and metabolic rate, but this is not so for resistance to cold. This shows that isofemale strains provide a quick way of assessing the possibility of associations between traits in natural populations, which can then be investigated for underlying causal relationships.

Desiccation has been studied in greater depth. A detailed consideration of the same two species for Brisbane (latitude $26°S$ and climatically intermediate between Melbourne and Townsville) and Melbourne populations clearly shows the possibility of using diallel crosses among isofemale strains to obtain an idea of the genetic basis of the trait (McKenzie and Parsons, 1974b) using methods outlined by Mather and Jinks (1971). The results are summarized in Table II and show

TABLE II. *Summary of diallel analyses for Melbourne and Brisbane populations indicating the number of significant components (P < 0.001) out of 8 for the Melbourne population and 4 for the Brisbane population (modified from McKenzie and Parsons, 1974b).*

Source of Variation[a]	Melbourne		Brisbane	
	melanogaster	simulans	melanogaster	simulans
a	8	8	4	4
b	8	0	4	4
c	0	0	0	0
d	0	0	0	0
Max. entry possible	8	8	4	4

[a] *a, tests primarily additive effects*
b, tests dominance effects
c. tests average maternal effect
d, tests the remainder of the reciprocal variation

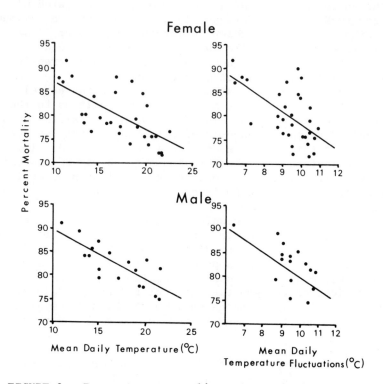

FIGURE 2. Percentage mortality of D. simulans from the Melbourne population for 12-h (males) and 16-h (females) desiccation periods graphed against mean temperature and temperature fluctuation with fitted regression lines (after McKenzie and Parsons, 1974b).

mortality resulting from desiccation to be under additive genetic control for all populations, with significant dominance for all but the Melbourne D. *simulans* population. Dominance, when present, is directional for resistance as is usually so for environmental stress traits (Parsons, 1973). Additionally when desiccation resistance was studied on an annual basis, only the Melbourne D. *simulans* population showed cyclical changes in mean mortality such that the population is most resistant in summer, becoming less so as the weather becomes cooler. This is shown in Figure 2, where mortality due to desiccation falls with increasing temperature and temperature fluctuation; these two variables are positively correlated in a temperate zone climate such as Melbourne. Adaptation to desiccation in this population is presumably by gene frequency changes, whereas in the other populations it is more likely to be a property of the entire genome involving interaction components. In other words, within the species *D. simulans*, the genetic basis of desiccation resistance differs between populations. The unit under consideration is quite clearly the population, not the species. It could well be that a feature of cosmopolitan species is that of a basic genome capable of adapting to a variety of environmental stresses by different means. Indeed, there is evidence that, for fruit-baited endemic Australian species, those species normally found only in rain forests are highly sensitive to desiccation and cold temperatures compared with the widespread cosmopolitan species *D. melanogaster* and *D. simulans* (Parsons and McDonald, unpublished data).

The ratio of *D. simulans* to *D. melanogaster* is higher in Brisbane than in Melbourne, temperature variability being lower in Brisbane. In general, *D. simulans* outnumbers *D. melanogaster* in regions where temperature fluctuations are small (McKenzie and Parsons, 1974a). Additionally, as a species *D. simulans* is more sensitive to temperature extremes and desiccation than *D. melanogaster*, so that Melbourne is a more marginal environment than Brisbane for *D. simulans*. The summer to winter desiccation resistance changes in Melbourne *D. simulans* are substantial (Figure 2) which argues for relatively few genes being involved. This rather "simpler" genetic basis for desiccation stress in Melbourne compared with Brisbane, may be analogous to the often observed reduced levels of chromosomal polymorphism at the margins of distributions of other *Drosophila* species (Carson, 1958). In Brisbane *D. simulans*, a highly integrated genome has presumably evolved including interaction components. In passing, it is worth suggesting that it would be of interest to look at the genetic basis of environmental stresses using the isofemale approach in well-studied species such a *D. pseudoobscura* where some, but not all, populations show seasonal cycles in inversion frequencies.

These examples indicate that it should be possible to obtain some understanding of the genotypes determining the distribution and abundance of *Drosophila* species from assessments of the genetic basis of resistance to these environmental stresses. As a prerequisite, a more precise knowledge of the effects of environmental heterogeneity on populations is necessary. For this extensive work, isofemale strains are valuable, especially as some information on the genetic basis of traits is obtainable irrespective of our knowledge of the genetics of the species, provided that a species can be cultured and strains crossed in the laboratory. Going beyond this stage to the study of actual genes involved requires a species such as *D. melanogaster*, where our genetic knowledge is extensive. However, the use of isofemale strains would appear to provide an informative starting point for the majority of species where genetic information is restricted. This would also appear to be applicable to other insects. A good example is the Queensland fruit fly, *Dacus tryoni*, in which there is demonstrated interpopulation variability for environmental stresses over a wide latitudinal range of eastern Australia. This is a species which has extended its geographical range in the last hundred years by means of physiological adaptation to extreme temperatures (Lewontin and Birch, 1966). It is, therefore, a species tending to become "more cosmopolitan" in its distribution, a phenomenon which apparently is also occurring for certain Australian endemic *Drosophila* species that come to fermented fruit baits (Parsons and Bock, 1977), although environmental stress studies on these have not yet been carried out.

III. RESOURCE UTILIZATION AND ALCOHOL

Given physical conditions of the environment permitting resource utilization, a series of related laboratory and field observations show that *D. melanogaster* can utilize an alcohol-associated resource, but *D. simulans* cannot (McKenzie and Parsons, 1972):

(1) *melanogaster* adults tolerate much higher ethanol concentrations in the environment than do *simulans*. Indeed, at certain concentrations likely to occur in nature, *melanogaster* longevity is increased (Starmer et al., 1977).
(2) a higher percentage of *melanogaster* larvae emerge from alcohol-containing media.
(3) gravid *melanogaster* females have a preference for alcohol-impregnated media, while *simulans* prefer standard media.
(4) only *melanogaster* occurs inside the wine-cellar at Chateau Tahbilk in Victoria at all life cycle stages, while both species are found outside.
(5) during vintage adult species ratios are affected such that the *melanogaster* frequency is high near the fermentation tank and decreases with distance from it, and release-recapture experiments show migration of *melanogaster* adults towards and *simulans* adults away from the fermentation tank at this time (McKenzie, 1974).

Based on 50 isofemale strains, larval tolerances to 9% alcohol were significantly correlated (McKenzie and McKechnie,

1978) with adult survival in vials saturated with ethanol fumes (r = 0.73, P < 0.001). Therefore, although larval survival is more relevant in considering resource utilization, it appears that all testable stages of the life cycle correspond in alcohol tolerances. In passing, this result brings up the possibility of using isofemale strains for looking at correspondences between life cycle stages for some of the other environmental stresses discussed.

Comparing populations within the wine-cellar, outside, and in Melbourne 100 km to the south, only *D. melanogaster* showed microdifferentiation in response to alcohol as a resource as expected; this occurred over a very short distance at Chateau Tahbilk and is describable in terms of adaptation to an alcohol resource (McKenzie and Parsons, 1974c). Plotting isofemale strain emergences from media supplemented with 9% alcohol shows this very clearly since sensitive strains found outside the wine-cellar were not found inside, leading to a shift in the mean (Figure 3). This shift occurs annually between vintages, since during vintage the cellar population is swamped by migration from outside, but at other times gene exchange is excluded leading to microdifferentiation (McKenzie, 1975b).

Alcohol tolerances in the Chateau Tahbilk population is determined polygenically as initially determined from a diallel cross among isofemale strains (McKenzie and Parsons, 1974c). Extreme strains were used for more detailed analyses to the chromosomal level; relatively major genetic effects were localized to chromosome II about 37 cM from the *Adh* locus and chromosome III near the *e* locus (McKenzie and McKechnie, 1978). At the winery the response strategy of the population to alcohol fumes is an increase in the frequency of tolerant alleles largely independent of the *Adh* locus, but this does not exclude the possibility of some functional relationship with the *Adh* locus (Ward, 1975). At first sight, however, this situation apparently contrasts with that of Briscoe *et al.* (1975) who found a higher Adh^F frequency in a Spanish wine cellar population than at a neighboring site and explained this by the relative ethanol resistance of adult *Adh* phenotypes. However, assuming alcohol to be a major selective force on the *D. melanogaster* genome, there is no reason all populations should respond similarly, since genes for alcohol resistance may differ among populations just as is so for resistance to desiccation. This could be looked at geographically, since adult tolerance to alcohol in the northern hemisphere increases with increasing latitude in *D. melanogaster* (David and Bocquet, 1975), but does not in *D. simulans*.

Indeed, the dissection of such geographical clines would be greatly aided by isofemale strains, whereby extreme strains

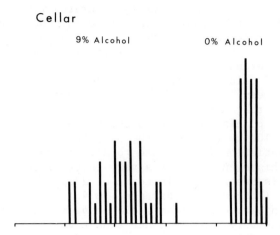

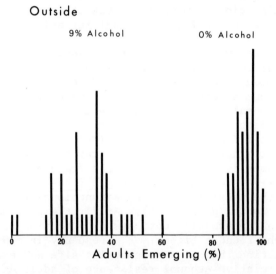

FIGURE 3. Distribution of the percentage of adults emerging from two replicates of 25 larvae each on 0% (control) and 9% alcohol-supplemented media for strains of D. melanogaster derived from inside and from outside the cellar at "Chateau Tahbilk". Each unit on the ordinate represents the mean of one strain. Cellar strains, mean emerging on 9% alcohol = 42.1±6.1; on 0% alcohol = 90.9±3.6. Outside strains, mean emerging on 9% alcohol = 28.5±9.2; on 0% alcohol = 91.6±4.3. (After McKenzie, 1975b, reprinted with permission).

TABLE III. Means and ranges of isofemale strains of D. melanogaster and D. simulans for number of larvae out of 10 choosing ethanol-containing agar after 15 minutes on a petri dish (after Parsons, 1977c).

	Tahbilk		Melbourne		Townsville	
	mean	range	mean	range	mean	range
D. melanogaster	7.5	2.3	7.8	1.8	6.4	5.2
D. simulans	5.3	1.6	5.5	1.5	5.8	1.8

from a population are isolated and studied in depth for the genetic basis of resistance just as has been done at Chateau Tahbilk. This possibility is illustrated by the behavior of newly hatched larvae in response to alcohol. As expected, southern Australian populations of D. melanogaster show a strong preference for alcohol-containing agar, while sympatric D. simulans populations show no such preference (Parsons, 1977c). This suggests that larvae seek out discrete and different microhabitats in nature in response to available resources, and incidentally suggests that the key to differences between some sibling species of Drosophila may well reside in differing larval resources utilized. Comparing the number of larvae choosing ethanol-containing agar out of 10 after 15 min on a petri dish, gave isofemale strain means and ranges for Melbourne, Chateau Tahbilk, and Townsville as in Table III. The means and ranges of the two southern D. melanogaster populations are similar, showing the expected alcohol preference, while all three D. simulans populations show the expected lack of preference with small ranges. In contrast, the Townsville D. melanogaster range is much larger, showing a high level of isofemale strain heterogeneity, such that some strains are within the range of the southern population, but others show almost no alchol preference. Presumably in southern Australia there is a greater premium on selection for alcohol resource exploitation than in the north. Indeed, the decline in heterogeneity in southern Drosophila agrees with the general principle of declining biological diversity with increasing latitude, demonstrated here for a chemically-defined resource rather than the more usual species diversity relationships (Emlen, 1973). The approach to the study of larval resource utilization via isofemale strain comparisons would appear to have considerable potential given that many possible metabolites are testable, and in any case the intrapopulation heterogeneity itself would have been difficult to study without considering variation among isofemale strains.

IV. AGING AMONG ISOFEMALE STRAINS WITH RESPECT TO ENVIRONMENT

Aging in natural populations has hardly been considered except in man where heredity and environment are almost impossible to disentangle. In *Drosophila*, most longevity studies have been carried out on inbred strains, their hybrids, and on mutants. One of the few studies on natural populations is that of Parsons (1977d, 1978b) on the longevity of 20 isofemale strains in *D. melanogaster* and *D. simulans* at each of the temperatures 20°C and 25°C. In *D. melanogaster*, differences among strains were found with a high correlation among isofemale strains across temperatures showing longevity to be under genetic control. In *D. simulans* there were differences among strains at each temperature associated with a large strains × temperature interaction such that the correlation among strains across temperatures was effectively zero. Quite clearly in this case, studies on the genetics of aging only provide information with respect to the environment chosen for study.

While the two environments are ostensibly similar for the two species, in fact they are not, since *D. melanogaster* is readily cultured at both temperatures while it is less easy to culture *D. simulans* at 25°C. This suggests that if the environment is sufficiently extreme to prevent species continuity, there may be some correspondences across environments. It is certainly not surprising that studies on the genetics of aging have not proceeded far, especially as both temperatures are encountered in nature although not, of course, annually. Both species will, in fact, survive in excess of 6 hours at 30°C without loss of fertility provided the humidity is high (Parsons, 1978c).

This rather surprising result gains support from work on ^{60}Co-γ irradiation in *D. melanogaster* where MacBean (1970) found no association between radiosensitivity to the extreme dose of 110 krads and longevity of untreated flies for 6 isofemale strains, and Westerman and Parsons (1972) obtained similar results for 4 inbred strains. Presumably there are differing genes or gene complexes determining longevity alone compared with longevity after γ-irradiation. The genetic basis of exposure to ^{60}Co-γ irradiation has been assessed by Westerman and Parsons (1973) at varying irradiation levels based on a diallel cross between 4 inbred strains (Table IV). At the extreme 120 krad stress, additive genetic control is

TABLE IV. *Levels of significance in D. melanogaster for general combining ability (G.c.a.) measuring additive genetic effects, and specific combining abilities (S.c.a.) measuring non-additive (dominance) effects together with the ratio of mean squares for G.c.a. and S.c.a. at exposures to ^{60}Co-γ irradiation between 0 and 120 krads. (Derived from data of Westerman and Parsons, 1973).*

	Dose (krads)					
	0	40	60	80	100	120
G.c.a.	<0.01				<0.05	<0.001
S.c.a.	<0.05	<0.001	<0.01	<0.001	<0.001	<0.001
G.c.a./S.c.a.	1.76	0.55	0.65	0.27	0.56	7.32

highly significant, and the non-additive component while significant at P < 0.001 is, in fact, minor by comparison. At all other doses, the non-additive (presumably dominance) effect is more important, while the control data show significant additive and non-additive effects, neither being large. Therefore, the genetic basis of longevity varies according to the level of environmental stress.

There is no reason why isofemale strains should not give similar results; indeed, a high level of additivity is obtained in the 100-120 krad range (Parsons et al., 1969). Given this level of additivity, gene localization studies based on Thoday's (1961) methods have been shown to be practicable using isofemale strains; indeed, such studies have been taken to the chromosome arm level in chromosomes II and III by MacBean (1970) and regions of major additive gene effects detected. The genetic basis of the polygenic trait, longevity, can therefore be most readily analyzed under extreme environmental stress conditions, such as the stresses already discussed as well as others, mainly of a chemical nature (Crow, 1957; Parsons, 1973).

V. DISCUSSION

Traits of ecological significance have been assessed mainly by using differential mortalities as the metric. In general, stresses are extremely severe with rapid death rates compared with unstressed flies. Under these circumstances, the genetic basis of traits is generally not complex, being highly additive and occasionally associated with dominance. Where death rates are not affected so drastically as in the aging experiments above, the heritability of longevity is lower, and our knowledge of its genetic basis rudimentary, because it is highly environment-dependent. In both cases isofemale strains play a role. Where the stresses are severe, an idea of the genetic basis of a trait is rapidly obtainable, while when less severe an estimation of the magnitude of the relative effects of genotype $v.$ environment in natural populations can be made.

Even though the high temperature/desiccation or low temperature extremes discussed are unlikely to affect populations continuously, the results correlate with annual population cycles as the comparative *D. melanogaster* - *D. simulans* data show. Indeed, in regions where such stresses become continuous, *Drosophila* will no longer be found. For example, the closely-related genus *Scaptomyza* occurs outside the *Drosophila* range both at the cold and hot extremes (Bock and Parsons, 1977). Within *Drosophila*, however, some species encounter very high temperatures in desert regions presumably by occupying moist microhabitats of cactus rot pockets, enabling high temperatures to be tolerated for a few hours at least (see Parsons, 1978a, for discussion). The same occurs for *D. hibisci* which utilizes the flowers of endemic *Hibiscus* species as a resource in open forests of the northern half of Australia where temperatures are frequently well over $30^{\circ}C$ (Cook *et al.*, 1977). In the laboratory it is not surprising that of the species tested for resistance to these extremes, *D. buzzati* is the most resistant. It is a species living in the rot pockets of the cactus *Opuntia* in open country where there would be a high premium on surviving extreme conditions. On the other hand, the most sensitive species so far tested, *D. paulistorum*, is from American tropical rain forests (Parsons and McDonald, unpublished).

Given the emphasis on polygenic variation of traits involved in population continuity, it is not surprising that behavioral traits as such have not been much considered, except as associated with other investigations, such as responses to alcohol. Even so, the importance of behavior should not be underestimated. For example, in adaptation to high temperatures there is the apparent tendency of flies to seek out moist microhabitats (Prince and Parsons, 1977). The evolutionary significance of many of the behaviors discussed in the literature is, however, most obscure (see Parsons, 1977e). There is an enormous literature on phototaxis, for example, with minimal considerations of variation in natural populations. However, Médioni (1962) found variations in populations of *D. melanogaster* at different localities in the northern hemisphere with positive phototaxis increasing going north. Under conditions of high temperature/desiccation stress, flies are likely to migrate to damp habitats which are often shaded, so that natural selection for negative phototaxis would be relatively high. Presumably such selection pressures would decrease with latitude, so making the observed phototaxis gradient plausible.

The most neglected behavioral traits concern larvae. However, Barker (1971) found that *D. simulans* larvae burrow deeper into media than *D. melanogaster*, which could be a behavioral adaptation to resource utilization if generalized across strains and populations. This, associated with interspecific differences in larval responses to alcohol, clearly indicates the usage of differing larval resources by the two species. Comparative studies of adult behaviors have not shown a great deal of divergence since differences between these two species are frequently quantitative, rather than qualitative (Parsons, 1975). Given that larvae utilize most of the resources needed for fly development, isofemale strain studies of sympatric populations appear potentially to be a valuable tool, especially if applied to a variety of potential resources. Such studies may aid us to determine more precisely what differentiates these two closely-related *Drosophila* species, and in due course other closely-related species. In this way, information on evolutionary divergence might be obtained. A considerable variety of metabolites may be involved in such studies, given that *Drosophila* specializes on a diversity of organisms causing fermentation and decay.

Biochemical variants, such as allozymes, have not been considered, though allelic differences at enzyme loci may form part of polygenic systems. A number of authors (*e.g.*, Johnson, 1974, 1976) have considered the possibility of biogeographic patterns of enzyme polymorphisms as a reflection of environmental factors as selective agents. In particular, latitudinal clines of polymorphic gene frequencies commonly occur in poikilotherms including *Drosophila*, suggesting that temperature may play an important role in selection for at least some of these polymorphisms (Clarke, 1975; Miller *et al.*, 1975), *e.g.*, alcohol dehydrogenase and α-glycerophosphate dehydrogenase. In addition, for substrates originating from the environment such as dietary proteins, alcohols, esters, and plant secondary proteins, considerable variations in available substrate type or concentration or both would be expected, perhaps on a latitudinal basis as suggested by larval reactions to alcohol already discussed. Associated allozyme frequency clines could well occur. Some of the clines and interpopulation differences discussed in this article, especially those relating to temperature, desiccation and alcohol, may in time be interpretable in more molecular terms than is presently possible. However, obtaining definitive data will be difficult if, following Johnson (1976), selection is envisaged as acting more upon integrated metabolic phenotypes than upon single loci -- a view which accords with Mather's (1943) balanced genotype, developed initially from a consideration of metrical traits of a morphological nature. The attempt is, however, worth making, especially as there is suggestive evidence for habitat selection associated with allozyme frequencies in *D. persimilis* from several ecologically distinct areas within a few hundred metres of each other (Taylor and Powell, 1977). Clarke (1975) and later Borowsky (1977) advocate the use of a correlation approach when investigating the possibilities of selection on protein polymorphism; as already discussed, correlations among isofemale strains may be informative for metrical traits, and the same must be true when considering protein polymorphisms in natural populations.

In conclusion, to understand polygenic variation in natural populations of *Drosophila* the variation must be conveniently and accurately assayed, as discussed here for traits of direct or inferred ecological significance. These are traits measured as proportions of a population surviving a stress, or proportions responding behaviorally to environmental heterogeneity in a population. Therefore, isofemale strains are characterized by proportions which vary at the intrapopulation level, and in total the proportions enable interpopulation comparisons. Extreme strains can then be used to study the genetic basis of traits within and between populations, as well as between species. Very few comprehensive studies have yet been carried out, but those few that have are encouraging. In genetically well-known species, of which there are few in the genus *Drosophila* other than *D. melanogaster*, specific genes controlling traits can be identified and studied. However, *a major advantage of isofemale strains is that genetic information is rapidly obtainable in species where details of the genome are unexplored*. This should surely aid in our assessment of the evolutionary significance of polygenic variation in natural populations, as has been shown here for *D. melanogaster*, and its sibling species *D. simulans* where the genome is less well-known.

VI. SUMMARY

The assessment of polygenic variation in natural populations of *Drosophila* is considered for acute environmental stresses relevant in determining population continuity, and behavioral responses to environmental heterogeneity. Genetic information is readily obtained from isofemale strains derived from natural populations, whether combined with a biometrical or non-biometrical genetic approach. In particular, genetic information is rapidly obtainable using a biometrical approach in species where details of the genome are unknown, which should aid in assessing the evolutionary significance of polygenic (and protein) variation in natural populations within and between species.

The isofemale approach is illustrated with comparative examples from the sibling species *D. melanogaster* where the genome is well-mapped and gene localization studies are possible, and *D. simulans* where the genome is less well-known. Examples considered include resistance to cold temperature, desiccation, and alcohol (including behavioral responses), and aging at two temperatures and various doses of ^{60}Co-γ irradiation. The use of isofemale strains in studying biogeographic variation is discussed.

ACKNOWLEDGMENTS

I am grateful to Dr. John A. McKenzie for a critical assessment of an earlier version of this manuscript, and the Australian Research Grants Committee for partial financial support.

REFERENCES

Andrewartha, H.G., and Birch, L.C. (1954). "The Distribution and Abundance of Animals." University of Chicago Press, Chicago.
Barker, J.S.F. (1971). *Oecologia* 29, 223-232.
Bock, I.R., and Parsons, P.A. (1977). *J. Biogeog.* 4, 327-332.
Borowsky, R. (1977). *Evolution* 31, 341-346.
Briscoe, D.A., Robertson, A., and Malpica, J-M. (1975). *Nature* 255, 148-149.
Carson, H.L. (1958). *Adv. Genet.* 9, 1-40.
Clarke, B. (1975). *Genetics* 79, 101-113.

Cook, R.M., Parsons, P.A., and Bock, I.R. (1977). *Aust. J. Zool.* 25, 755-763.
Crow, J.F. (1957). *Ann. Rev. Entomol.* 2, 227-246.
David, J., and Bocquet, C. (1975). *Nature* 257, 588-590.
Dobzhansky, Th. (1950). *Amer. Scientist* 38, 208-221.
Dubinin, N.P., and Tiniakov, G.G. (1947). *Am. Nat.* 81, 148-153.
Emlen, J.T. (1973). "Ecology: An Evolutionary Approach." Addison-Wesley, Reading, Massachusetts.
Fisher, R.A. (1918). *Trans. Roy. Soc., Edin.* 52, 399-433.
Johnson, G.B. (1974). *Science* 184, 28-37.
Johnson, G.B. (1976). *In* "Molecular Evolution" (F.J. Ayala, ed.). Sinauer, Sunderland, Massachusetts.
Lee, B.T.O., and Parsons, P.A. (1968). *Biol. Reviews* 43, 139-174.
Lewontin, R.C., and Birch, L.C. (1966). *Evolution* 20, 315-336.
MacBean, I.T. (1970). Ph.D. Thesis, La Trobe Univeristy, Bundoora, Victoria, Australia.
Mather, K. (1943). *Biol. Reviews* 18, 32-64.
Mather, K., and Harrison, B.J. (1949). *Heredity* 3, 1-52, 131-162.
Mather, K., and Jinks, J.L. (1971). "Biometrical Genetics." Chapman and Hall, London.
McKenzie, J.A. (1974). *Oecologia* 15, 1-16.
McKenzie, J.A. (1975a). *Aust. J. Zool.* 23, 237-257.
McKenzie, J.A. (1975b). *Genetics* 80, 349-361.
McKenzie, J.A., and McKechnie, S.W. (1978). *Nature* 272, 75-76.
McKenzie, J.A., and Parsons, P.A. (1972). *Oecologia* 10, 373-388.
McKenzie, J.A., and Parsons, P.A. (1974a). *Aust. J. Zool.* 22, 175-187.
McKenzie, J.A., and Parsons, P.A. (1974b). *Aust. J. Biol.* 27, 441-456.
McKenzie, J.A., and Parsons, P.A. (1974c). *Genetics* 77, 385-394.
Médioni, J. (1962). *Bull. Psychol., Paris* 16 (2), 8.
Miller, S., Pearcy, R.W., and Berger, E. (1975). *Biochem. Genet.* 13, 175-188.
Parsons, P.A. (1973). *Ann. Rev. Genet.* 7, 239-265.
Parsons, P.A. (1975). *Genetics* 79, 127-136.
Parsons, P.A. (1977a). *Amer. Nat.* 111, 613-621.
Parsons, P.A. (1977b). *Aust. J. Zool.* 25, 693-698.
Parsons, P.A. (1977c). *Oecologia* 30, 141-146.
Parsons, P.A. (1977d). *Exp. Geront.* 12, 241-244.
Parsons, P.A. (1977e). *Adv. Genet.* 19, 1-32.
Parsons, P.A. (1978a). *Amer. Nat.*, in press.
Parsons, P.A. (1978b). *Exp. Geront.* 13, 167-169.
Parsons, P.A. (1978c). *Genetics* 88, s77.

Parsons, P.A., and Bock, I.R. (1977). *Aust. J. Zool.* 25, 249-268.
Parsons, P.A., MacBean, I.T., and Lee, B.T.O. (1969). *Genetics* 61, 211-218.
Payne, F. (1918). *Indiana Univ. Studies 5 (36)*, 1-45.
Prince, G.J., and Parsons, P.A. (1977). *Aust. J. Zool.* 25, 285-290.
Starmer, W.T., Heed, W.B., and Rockwood-Sluss, E.S. (1977). *Proc. Natl. Acad. Sci., U.S.A.* 74, 387-391.
Tantawy, A.O., and Mallah, G.S. (1961). *Evolution* 15, 1-14.
Taylor, C.E., and Powell, J.R. (1977). *Genetics* 85, 681-695.
Thoday, J.M. (1961). *Nature* 191, 368-370.
Thoday, J.M. (1967). *Ciencia e Cultura* 19, 54-63.
Ward, R.D. (1975). *Genet. Res.* 26, 81-93.
Westerman, J.M., and Parsons, P.A. (1972). *Int. J. Radiat. Biol.* 21, 145-152.
Westerman, J.M., and Parsons, P.A. (1973). *Can. J. Genet. Cytol.* 15, 289-298.
Wiens, J.A. (1977). *Amer. Scientist* 65, 590-597.

*The Biometrical Approach
to Quantitative Variation*

THE BIOMETRICAL APPROACH TO QUANTITATIVE VARIATION

J. L. Jinks

Department of Genetics
University of Birmingham
Birmingham, U.K.

I. INTRODUCTION

Among the many methods described in this volume for studying quantitative variation, biometrical genetics (Mather, 1949; Mather and Jinks, 1971, 1977; chapters by Mather and by Caten in this volume) stands alone in offering a comprehensive and coordinated set of procedures that make it as universal in its applications as is the Mendelian method for studying qualitatively inherited differences. Unlike the latter, however, biometrical genetics can accommodate all of the phenotypic variation displayed by a character irrespective of its properties or its source. And because it is firmly based on Mendelian principles, it retains the interpretive and predictive powers of Mendelian genetics, but only in respect of the summed or net effect of all gene loci that are contributing to the variation.

There is, of course, another powerful technique for analysing quantitative variation, namely, chromosome assay (see chapters by Mather and by Oliverio in this volume), but it, unlike biometrical genetics, can only deal with very limited parts of the variation except in two or three intensively studied species and is never likely to be much more widely applicable. There are also analyses, for example, those based upon response to selection (see chapter by Geldermann and Gundel and others in this volume) that are as widely applicable as biometrical genetics, but they permit only the most superficial partitioning of the variation among the various heritable and non-heritable sources. Analyses based solely on the concepts and procedures of statistics are equally

widely available and can often be very powerful, but unless
the components of the statistical model are directly equatable,
without making crippling assumptions, to components of a genetical model -- which they never are unless the most sophisticated biometrical genetical experimental designs are used on
material of known genetical constitution (section VIII) -- the
components have only limited value in interpreting the underlying gene action or for making predictions across generations
(see for example, Griffing, 1956; Gilbert, 1967). Hence in
most statistical analyses, the handling of the heritable variation is no less empirical than that of the non-heritable
variation and little or no use is made of the analytical tools
of Mendelian genetics for which there is no substitute for un-
ravelling heritable variation.

The ultimate aim of a biometrical genetical analysis is to
partition the total phenotypic variation among heritable and
non-heritable components such that each component is the contribution of one source only to one of the statistics: mean,
variance, skewness and kurtosis, that describe the variation.
Since the properties in transmission and expression of each
source of heritable variation is known from the study of major
genes, estimates of these components provide all the information that is required to interpret the heritable variation and
to predict how it will behave in future generations. The many
and varied applications of biometrical genetics all stem from
these properties of the components.

Biometrical genetics sets out to achieve its aims through
three closely inter-dependent approaches. (i) The definition
of the components of means, variances, skewness and kurtosis
and the derivation of theoretical expectations which define
their contributions to each statistic for every type of family,
generation, or population of interest (section IV). (ii) the
development and testing of experimental designs and breeding
programmes with the aim of increasing the amount of information that can be obtained about estimates of these components
(section VII), and (iii) the development of procedures for
processing and analysing the data to obtain the best estimates
of the components and from them drawing conclusions and making
predictions (section VIII). The success of the latter is,
of course, dependent on the definition of the components, the
validity of the theoretical expectations and the quality of
experimental data. Thus, advances in either the theory or
practice of biometrical genetics inevitably exposes limitations and the need for further developments in the other.

II. SOURCES OF VARIATION

For twenty years or more biometrical genetical models of quantitative or continuous variation have been available that explicitly incorporate all the sources of variation that have been identified from the study of discontinuous variation. The contributions these sources make to the total phenotypic variation in haploid, diploid, and polyploid species have been generalised to accommodate any number of gene loci and environmental factors and all possible interactions among allelic and non-allelic genes and between these and the environmental factors. A brief description of these sources and their associated symbols follows: (for a fuller account see Mather and Jinks, 1971, 1977, and for further information see chapters by Mather and by Caten in this volume).

(1) Additive genetic component; arises from the differences between a pair of corresponding homozygotes. Symbolised by d, the departure of one of the homozygotes from the origin m, which is positive for the homozygote carrying the increasing allele (A) and negative for that carrying the decreasing allele (a).

(2) Dominance component; arises from the departure of the heterozygote from the mean of the corresponding pair of homozygotes. Symbolised by h, which is positive when the heterozygote is more like the higher scoring of the two corresponding homozygotes and negative when it is more like the lower scoring.

(3) Additive environmental component; arises from the average effect of the environment on all genotypes. Symbolised by e and measured as the average departure of all genotypes from the origin that is ascribable to environmental causes.

(4) Non-allelic interactions; arise from modification of the additive and dominance effects at one locus resulting from allelic substitution at other loci. For pairs of loci these may be of three kinds, symbolised i, j, and l; i symbolises the modification of the additive effect at one locus resulting from a substitution of one homozygous state for the alternative homozygous state at a second locus; j symbolises the modification of the dominance effect at one locus when one homozygous state is substituted for the alternative homozygous state at a second locus and l symbolises the modification of the dominance effect at one locus ascribable to heterozygosity at a second locus. These definitions and symbols are readily extendable to interactions among three or more loci (Mather and Jinks, 1971).

(5) Genotype × environment interaction; arises from differences between genotypes in departures from the origin ascribable to environmental causes. Symbolised by g with subscripts d, h, i, j, l, *etc.* according to the kind of gene action or interaction that is involved in the interaction with the environment.

(6) Differences between reciprocal crosses; is self-explanatory and arises from two major causes, sex-linkage and maternal or more rarely paternal effects. Neither requires additional symbols since their effects can be expressed in terms of the additive, dominance, and interaction effects of genes which are either sex-linked or whose expression has been delayed for one or more generations. Subscripts added to the normal symbols for gene action and interaction are sufficient to identify them as arising from genes which are sex-linked or part of the maternal (paternal) genotype.

The origin m, from which all effects are defined as departures, is the mean of all possible homozygotes, in respect of all the loci at which there are allelic differences, when raised in equal frequencies in all relevant environments. In the absence of both non-allelic interactions and genotype × environment interaction, this origin is equivalent to the mean of the real or notional pair of pure-breeding homozygotes from which all the genotypes under consideration have either arisen or from which they could have arisen following a cross between them. Providing that all genotypes under consideration are raised in a randomised layout within a single environment, the additive environmental effect, which in these circumstances is the only effect ascribable to the environment, becomes part of the origin, m.

The sources of variation have been defined without reference to such important concepts of classical genetics as autosomal linkage and the breeding system. This is because factors such as these, which determine gene and genotype frequencies and linkage disequilibria, make no direct contribution to variation; their role is the determination of the relative contributions that each of the heritable sources of variation make to the total variation; they determine the composition of the variation without themselves being sources of variation.

III. SCALING TESTS

Although all sources of variation have been incorporated individually and simultaneously into the models and procedures of biometrical genetics, in practice experimental data are rarely so extensive that the majority, let alone all, can be followed in a single experiment or even in a number of interrelated experiments. This, however, is not necessarily a disadvantage, since it is unlikely that all sources will be contributing at a statistically significant level to the variation observed in any one study. Furthermore, the analytical procedures achieve their greatest efficiency when the sources of variation that are being investigated simultaneously are the minimum that can adequately account for the observed variation. The importance of establishing at an early stage in a biometrical genetical analysis which of the sources are contributing at a statistically significant level has long been recognised. To this end statistically simple, reliable tests, usually referred to as scaling tests, have been devised (Mather, 1949; Mather and Jinks, 1971, 1977).

The earliest scaling tests were used to detect the presence of interactions between non-allelic genes and between genotype and environment, neither of which could be handled by the biometrical genetical procedures available at that time (Mather, 1949). There was a need, therefore, to test the null hypothesis that neither was contributing significantly to the variation so that in the event of the failure of the hypothesis no analysis or interpretation would be attempted which overlooked their presence. Until, however, the methodology for specifying and analysing their contributions had been devised (Hayman and Mather, 1955; Mather and Jones, 1958) there were severe restrictions on further analysis if such a failure occurred. The only logical way forward was to seek a scalar transformation of the data that would minimise these interactions (Mather, 1949).

Scaling tests are now available for detecting every conceivable source or combination of sources of variation and, hence, for testing the validity of the assumption that they are absent. However, it is no longer necessary that the more complex sources of variation should be absent in order to proceed with a satisfactory biometrical genetical analysis. Not only can their contributions be incorporated into the theoretical models and expectations, but the magnitude of their contributions can be estimated from data with an appropriate structure (Mather and Jinks, 1971, 1977; Jinks and Perkins, 1969; Perkins and Jinks, 1970).

The essence of a scaling test is a simple relationship between the means or variances of related families which holds only if a specific source of variation is absent. Powerful statistical tests can then be used to see if the relationship holds and, hence, to test the null hypothesis that the source is not contributing at a significant level. Such tests have been developed to detect specifically and unambiguously contributions from digenic and higer order non-allelic interactions (Mather, 1949; Jinks and Jones, 1958; Mather and Jinks, 1971, 1977), trigenic and higher order non-allelic interactions (Van der Veen, 1959; Jinks and Perkins, 1969; Mather and Jinks, 1971), genotype × environment interactions (Mather, 1949; Bucio Alanis, Perkins, and Jinks, 1969; Perkins and Jinks, 1971), sex-linkage (Mather and Jinks, 1963, 1971; Killick, 1971), maternal and grandmaternal effects (Mather and Jinks, 1963, 1971; Jinks *et al.*, 1972), autosomal linkage (Perkins and Jinks, 1970), linkage of genes which are displaying non-allelic interactions (Jinks, 1978), genotype-environment correlation and cultural transmission (Eaves, 1976; Eaves, Last, Martin, and Jinks, 1977). The range of tests available depends on the nature of the breeding material and the design of the breeding programme (section VII). In general the most recently introduced designs, *e.g.* the Triple Test Cross, provide the most comprehensive set of tests (Kearsey and Jinks, 1968; Jinks and Perkins, 1970; Perkins and Jinks, 1970, 1971). So by a judicious choice of design, detection of the more complex sources of variation can precede the process of estimating the contributions that the significant sources are making to the observed variation. Ideally, therefore, the experimental data alone can determine the level of complexity of the biometrical genetical analysis required.

IV. STATISTICS, PARAMETERS, AND MODELS

To describe completely the distribution for a continuously varying character within a family, a generation or a population we must know the mean and variance of the distribution and in some circumstances its skewness and kurtosis also. We, therefore, require to know the expectations of these statistics in terms of the sources of variation that are contributing to them for each type of family, generation and population that could be encountered in controlled breeding programmes or in surveys of naturally breeding populations. The contributions of the heritable sources of variation are determined solely by gene and genotype frequencies and those of the non-heritable sources by the experimental design and in particular the unit of randomisation. To prodeed further with the definition of these contributions, however, we must introduce the concept of balance (Mather, 1941). Balance has no counterpart in classical genetics, but its introduction is a prerequisite for understanding heritable components which describe the contributions of genes at two or more loci with similar and supplementary effects on the phenotype.

A. *Means*

Let us consider two pure-breeding lines (P_1 and P_2) of a diploid species that are homozygous for different alleles at k loci each of which is contributing an additive effect d_i where i = 1 to k to their mean phenotypes $\overline{P}_1$ and $\overline{P}_2$, respectively. If P_1 carries the increasing allele and hence P_2 the decreasing allele at all k loci, that is, like alleles are associated in the parents, the contribution of the additive genetic component to the mean of P_1 will be $+ \sum_{i=1}^{k} d_i$ and to P_2, $- \sum_{i=1}^{k} d_i$ and $\overline{P}_1$ and $\overline{P}_2$ will differ by $2 \sum_{i=1}^{k} d_i$. In these circumstances the difference between $\overline{P}_1$ and $\overline{P}_2$ will reflect the total additive genetic differences between them. In general, however, P_1 while having a larger mean (by definition) will carry the decreasing allele at some loci (at k' of the k loci) and, hence, P_2 the increasing allele at those same loci so that the contribution of the additive genetic component to $\overline{P}_1$ will be $\sum_{i=1}^{k} d_i - 2\sum_{i=1}^{k'} d_i$, which is positive and to $\overline{P}_2$, $-\sum_{i=1}^{k} d_i + 2\sum_{i=1}^{k'} d_i$, which is negative. Consequently the difference between $\overline{P}_1$ and $\overline{P}_2$, that is the additive genetic difference we can observe, is $2 \sum_{i=1}^{k} d_i - 4\sum_{i=1}^{k'} d_i$. This is, of course, less than the total additive genetical differences at the k loci at which they differ. This balancing out between loci making positive and negative contributions is a general property of the genetical components of family, generation and population means. The cause in this particular instance is that like alleles will, in general, be partially dispersed rather than completely associated in a pair of pure-breeding lines. But it can also arise from another cause. For example, the F_1 of a cross between P_1 and P_2 will be heterozygous at all k loci and each locus will contribute a dominance deviation h_i to the mean of the F_1. If the increasing allele at the ith locus is dominant h_i will be positive, but if the decreasing allele is dominant it will be negative. Unless, therefore, the dominance deviations at all k loci are the same sign, that is, dominance is unidirectional, $\sum_{i=1}^{k} h_i$ will be the net balance of positive and negative values and the dominance component we can observe as a deviation of the F_1 from the origin, m, will be less than the total dominance deviations at all loci. The theoretical expectations for means must, therefore, be expressed in terms of the net directional contributions of those sources of variation, which like dominance and non-allelic interactions, can take sign. This clearly imposes restraints on their interpretation (section VI).

Once it is recognised that the components of means, symbolised as [d], [h], [i], [j], [l], *etc.*, refer to net directional effects, the specification of the expected means can follow directly from considerations of gene and genotype frequencies at single loci. This may be illustrated by considering the families produced by selfing, backcrossing, or sib-mating following an initial F_1 cross between a pair of pure-breeding lines of a diploid species. In respect of any one locus, A,a, the F_1 will have the genotype Aa and this locus will contribute the dominance deviation h_a to the mean. For k such loci the contribution to the F_1 mean will be [h], the net directional dominance component. For any pair of loci, A,a; B,b, the F_1 will have the genotype AaBb and this pair of loci will contribute the digenic non-allelic interaction deviation l_{ab} to the mean.

There are $\frac{1}{2}k(k-1)$ such pairs of loci each of which can contribute an l to the F_1 mean and each l may be either positive or negative depending on the type of interaction (section VI). Their overall contribution to the F_1 will, therefore, be the net directional balance symbolised by [l]. Hence, the expected F_1 mean assuming no higher order interactions is

$$\overline{F}_1 = m + [h] + [l].$$

In a selfing series initiated from this F_1, the frequency of heterozygotes will fall at a predictable rate such that in the nth generation (F_n) their frequency will be $(\frac{1}{2})^{n-1}$. In the absence of linkage, the frequency of simultaneous heterozygosity at pairs of loci will be the frequency of heterozygosity raised to the power of 2. The expected mean for the F_n generation, therefore, is

$$\overline{F}_n = m + (\tfrac{1}{2})^{n-1} [h] + (\tfrac{1}{2})^{2(n-1)} [l].$$

In the absence of selection all generations derived from an initial cross between two pure-breeding lines by selfing, sib-mating, and random mating have equal frequencies of increasing and decreasing alleles at every one of the k loci. In the expected means of these generations there are, therefore, no deviations from m ascribable to the additive effects of the genes (d) or to interactions involving additive effects (i and j). Generalised, the expected means of these generations (Ȳ) will, therefore, always be of the form

$$\bar{Y} = m + x[h] + x^2[l_{12}] + x^3[l_{123}] + \ldots x^k[l_{1\ldots k}]$$

where x is the expected frequency of heterozygosity at a single locus in that generation and $[l_{12}]$, $[l_{123}] \ldots [l_{1\ldots k}]$ are the net directional effects of interactions among heterozygotes at pairs of loci, at loci taken three at a time, and so on, up to at all k loci taken simultaneously.

If the gene frequencies are unequal there are contributions to the generation means arising from the additive effects of the genes with coefficients determined by the inequality in gene frequencies. For example, in the B_1 generation produced by backcrossing F_1 to P_1 the frequency of alleles from the P_1 parent is 3/4 and from the P_2 parent, 1/4. The contribution of the net additive effect, $[d]$ to the mean of B_1 is, therefore, $3/4[d] - 1/4[d] = 1/2[d]$. Since the frequency of heterozygosity in B_1 is 1/2 the contribution of the net directional dominance component $[h]$ is $1/2[h]$. In the absence of interactions, therefore,

$$\bar{B}_1 = m + 1/2[d] + 1/2[h].$$

Extended to include the digenic interactions it becomes

$$\bar{B}_1 = m + 1/2[d] + 1/2[h] + (1/2)^2[i] + (1/2)(1/2)[j] + (1/2)^2[l]$$

where the coefficient of $[i]$ is the square of the coefficient of $[d]$ and the coefficient of $[j]$ is the product of the coefficients of $[d]$ and $[h]$.

Generalised, the expected mean of a generation with unequal gene frequencies is

$$\bar{Y} = m \pm y[d] + x[h] + y^2[i_{12}] \pm xy[j_{12}] + x^2[l_{12}]$$
$$\pm y^3[i_{123}] \pm x^2y[j_{12/3}] + xy^2[j_{1/23}] + x^3[l_{123}], \text{etc.}$$

where y is the inequality in gene frequencies and $j_{12/3}$ and $j_{1/23}$ are interactions between a pair of homozygous loci and a heterozygous locus and between a homozygous locus and a pair of heterozygous loci, respectively. The upper sign applies if the alleles from P_1 are in the majority and the lower sign if the reverse holds.

The general principle is, therefore, quite simple; the coefficients of [d] and [h] are determined solely by the expected gene frequencies and genotype frequencies at a single locus and, in the absence of linkage, the coefficients of the digenic, trigenic, and higher order interactions can be derived as appropriate powers and products of the coefficients of [d] and [h]. Extension of these expectations to include genotype × environment interactions, sex-linkage and maternal effects and their modification for haploid and polyploid species introduces no additional rules (Mather and Jinks, 1971). Autosomal linkage never influences the coefficients of [d] and [h] because linkage does not alter gene and genotype frequencies at single loci. It does, however, alter the coefficients of the interaction components which also now depend upon the relative frequencies of coupling and repulsion phases (Jinks and Perkins, 1969; Mather and Jinks, 1971; Jinks, 1978).

B. *Variances*

Unlike means, variances reflect the total effects of the contributory sources of variation. Hence, the informational content of a variance is both statistically and genetically different from, and complementary to, that of the mean. There is no sense, therefore, in which an analysis of means or of variances can be regarded as alternatives as has sometimes been suggested.

The components in terms of which contributions to the expected variances are expressed take the form

$$D = \sum_{i=1}^{k} d_i^2, \quad H = \sum_{i=1}^{k} h_i^2, \quad \text{and} \quad E = \sum_{j=1}^{t} e_j^2.$$

These are the additive genetic, dominance, and additive environmental components of variation for equal gene frequencies at all k loci (see also chapter by Mather in the volume). Each component is clearly the total contribution of each source of variation. The contributions that each of the genetic components makes to the variation in any particular family, generation, or population follows directly from the expected gene and genotype frequencies at a single locus. For the generations that can be derived by selfing, sib-mating, or random mating from an initial F_1 cross between a pair of pure-breeding lines of a diploid species the variance (V) always takes the form

$$V = (1-x)D + x(1-x)H + E$$

where x is the expected frequency of heterozygotes and $(1-x)$ the frequency of homozygotes at each of the k loci.

With unequal but known gene frequencies such as would result from backcrossing to P_1 or P_2 a fourth component

$$F = \sum_{i=1}^{k} d_i h_i$$

is required and the generalised variance becomes

$$V = y(1-y)D + x(1-x)H \mp 2xyF + E$$

where y is the difference in frequency of alleles that originated from P_1 as opposed to P_2 and the upper sign of F applies when the alleles from the P_1 parent are in excess and the lower sign when the reverse holds.

These expectations, like those for the means, can be extended to accommodate common family environments, non-allelic interactions, genotype × environment interactions, maternal effects, sex-linkage, and autosomal linkage, and be modified for haploid and polyploid species (Mather and Jinks, 1971).

There are few experimental designs (section VII) that will not permit the total variation within a population or generation to be partitioned between two or more variances on the basis of an internal structure such as two or more degrees of relationship among its members. The simplest structure is where all individuals belong to families of a single kind, for example, unrelated families of full-sibs. The total variation can then be divided into two parts, the variance of family means (which is the full-sib covariance in this example) and the mean variance within families between individuals. The expectations of both can then be given in terms of the heritable and non-heritable components of variation which must sum to the expectation of the total variance. The more items that the total variance can be partitioned into in this way the more informative the subsequent biometrical genetical analysis. And as we shall see (section VII) considerable ingenuity in experimental design has been exercised to achieve the maximum partitioning of the total variance within the context of an analysis of variance and to separate out the contributions of different sources of variation into different mean squares. In this way testing for a significant contribution from a particular source in an analysis of variance and estimating the magnitude of its contribution as a component of an expected mean square are successive stages in the same analysis.

Partitioning of the total variation within generations derived from an initial cross between a pair of pure-breeding lines on the basis of an internal hierarchical structure has long been a standard procedure in a biometrical genetical analysis (Mather, 1949). In the nth generation there are n-1 possible hierarchies or ranks (r=1 to n-1) based upon common parents (r = n-1), grandparents (r = n-2), great grandparents (r = n-3) *etc.*, so that relatively early in, for example, a selfing (F) series each generation is yielding a number of independent mean squares among which the total variation may be partitioned. The expected contributions of the heritable components, D and H, to the corresponding variances for the F series are

$$V_{rFn} = (1/2)^r D + (1/2)^{2n-r-1} H$$

and the non-heritable components as usual depend on the scheme of randomisation. Perhaps more importantly, however, within any one series, for example the selfing or backcrossing series, the effect of autosomal linkage on the heritable components is the same for all variances of the same rank (same r) irrespective of the generation (n) (Mather, 1949; Mather and Jinks, 1971). This provides the basis of all methods of detecting linkage disequilibria and classifying its predominant phase, coupling or repulsion.

V. MODEL FITTING

There are two stages to the estimation of the contribution of the various sources to the phenotypic variation for any character. The first is to recognise by appropriate tests of significance which sources are contributing at a significant level. The second is to evaluate the magnitude of the contributions from the significant sources using one of a number of standard model fitting procedures. Ideally one should aim to have at least one more observed statistic than the number of components that have to be included in the model in order to accommodate all the significant sources of variation. This then permits a test of goodness of fit of the model using maximum likelihood or weighted least squares estimates of the components as well as tests of significance of the individual components (Cavalli, 1952; Jinks, 1956; Mather and Jinks, 1971, 1977; Gale, Mather, and Jinks, 1977).

A satisfactory model is then one in which the test of goodness of fit of the observed and expected statistics is non-significant, all the individual components of the model are significant, and the magnitudes and signs of the components are statistically and genetically acceptable and sensible. In theory the components of such a model should be those which correspond with the sources of variation which were found to be making a significant contribution in the preliminary scaling tests. In practice this is almost invariably the case when fitting models to means but it is regularly so only for the simpler models when fitted to variances. For example, either linkage or non-allelic interactions can be readily detected in the presence of the other by scaling tests applied to the variances whereas demonstrating their presence by fitting models to variances can be inconclusive if both are present (Perkins and Jinks, 1970; Opsahl, 1956).

An important generalisation that should be borne in mind when fitting any genetical model is that most of the heritable differences in phenotype will generally, though not, of course, necessarily, be accounted for by the additive and dominance components and most of the rest by one of the first order interactions, *e.g.* digenic interactions, genotype × environment interactions involving additive and dominance gene action only, and linkage of non-interacting genes. Unless, therefore, a scaling test unambiguously indicates otherwise, there is no point in considering more complex situations until a simpler model has been fully explored and has clearly failed to provide an adequate representation of the variation in phenotype.

VI. INTERPRETATION

Statistically sound estimates of the significant components of a satisfactory model are the starting point of the third stage of a biometrical genetical study, the interpretation. This is the stage where the interdependence of the information from means and variances becomes most apparent and the relationships between the components of the model and the concepts of gene action and interaction gained from classical genetics become most crucial.

As we have already observed, the relative magnitudes of the different kinds of gene action and interaction cannot be inferred from estimates of the components of the means (section IV). A relatively small value of $[d]$ may imply little or no additive genetic differences between the initial parents of a cross or in the generations that can be derived from it.

But unless these parents were extreme selections, it is more likely that there is a somewhat larger additive genetic difference than the estimate of $[d]$ would imply with dispersion of increasing and decreasing alleles between them (Jayasekara and Jinks, 1976). Equally a relatively small value of $[h]$ may imply either little or no dominance or ambidirectional dominance. In general, therefore, the relative values of $[d]$ and $[h]$ can tell us little about the dominance properties at the k loci. The most that can be safely inferred from them is whether the dominance has any net direction and if so, whether increasing or decreasing alleles are more frequently the dominant alleles.

Exactly the same problem arises with the digenic non-allelic interaction components $[i]$, $[j]$, and $[l]$. At the same level of individual pairs of loci the d, h, i, j, and l parameters are defined such that all the classical non-allelic interactions can be identified from their relative magnitudes and signs (Hayman and Mather, 1955; Mather and Jinks, 1971, 1977). This does not, however, extend to the components that are estimated, namely, $[d]$, $[h]$, $[i]$, $[j]$, and $[l]$. Of these, only the magnitudes and signs of $[h]$ and $[l]$ are uninfluenced by the way the alleles are distributed between the parents of the cross which initiated the breeding programme. Only $[h]$ and $[l]$, therefore, reflect the net direction of the gene action and interaction and hence only these two components can be used to classify the predominant type of epistasis (Jinks and Jones, 1958; Mather and Jinks, 1971, 1977). As a consequence only two broad classes of non-allelic interactions can be recognised; in one $[h]$ and $[l]$ have the same sign which indicates that the interactions are predominantly of a complementary kind and in the other $[h]$ and $[l]$ have opposite signs indicating that they are predominantly of a duplicate kind. In all, therefore, four possibilities can be recognised since for both kinds of interactions the net direction of the dominance component $[h]$ may be positive or negative according to whether the increasing or the decreasing allele is more often the dominant allele.

The estimate of $[i]$ is the most seriously affected by the dispersion of like alleles. If more than half of the pairs of interacting loci have like alleles dispersed, the sign of $[i]$ will be the reverse of that of the sum of the i's over all pairs of interacting loci. While, therefore, i and l will always have the same sign for each pair of loci showing any of the classical interactions, $[i]$ and $[l]$ will have opposite signs if at these interacting loci like alleles are predominantly dispersed (Jinks and Jones, 1958).

These same considerations apply equally to the interpretation of the other components of means specifying the effects of higher order interactions, genotype × interactions, sex linkage, and maternal effects (Mather and Jinks, 1971, 1977).

With so many restrictions on their interpretation one might well ask whether estimating the components of mean serves any useful purpose. The position can be summarised as follows:

    1. They can usually be readily and reliably estimated from relatively modest experiments and the estimates provide a direct demonstration of the sources of variation which have a net directional contribution.
    2. They can be used to predict the means of future generations and in conjunction with the corresponding components of the variances, their relative distributions (section VIII B).
    3. In analysing the genetical architecture of a character the net directional effects are often as informative as the total effect (section VIII A).

Nevertheless, an analysis confined to means leaves many questions unanswered that are readily answered by the components of the variances. For example, the relative values of $[d]$ and $[h]$ tell us nothing about even the average dominance properties at the k loci; the ratio $\sqrt{H/D}$ on the other hand is a weighted average dominance ratio for all k loci, but tells us nothing about the direction of the dominance. By combining the information on the sign of $[h]$, the value of $\sqrt{H/D}$, the sign of F and the value of $F/(\sqrt{D \cdot H})$, one can infer whether the increasing or decreasing alleles are more often the dominant allele, the average dominance ratio and its variation over loci and the proportion of dominant to recessive alleles in each of the initial parents of the cross.

A similar situation arises over autosomal linkage where only the components of the variances can be used to detect linkage of non-interacting gene loci and to determine its predominant phase, *i.e.* coupling or repulsion. However, linkage of gene loci displaying non-allelic interactions is more readily detected and classified by using the components of means (Jinks and Perkins, 1969; Jinks, 1978). Again no analysis of the variation attributable to environmental causes or to genotype × environment interaction, other than that arising from deliberately imposed macro-environmental treatments, is possible without recourse to the components of the variances. Where genotypes and families are replicated in different environments the analysis may proceed at the level of both means and variances, but in general the analysis of means can usually be pursued further and is the more rewarding (Bucio Alanis *et al.*, 1969; Perkins and Jinks, 1970).

These examples illustrate how the advantages and disadvantages of pursuing an analysis at the level of means or of variances is conditioned by a number of factors including the nature of the data, the aims of the enquiry and the complexity of the genetical control encountered. There is no theoretical limit to the complexity that can be handled at the level of variances, practical considerations, however, of which the dimensions of the data are paramount, usually reduce the resolving power of the analysis of variances well below that which is achievable with means. But as we shall see in section VIII, the solution of many problems requires reliable and simultaneous information about components of means and of variances and it is, therefore, necessary to design experiments that will provide both (section VII).

VII. EXPERIMENTAL DESIGN

It almost goes without saying that the quantitative data used in a biometrical genetical analysis should have been obtained from an experimental design that satisfies all the normal statistical criteria. Over and above these, however, there are additional requirements relating to the composition of the material which are peculiar to biometrical genetics, and attention to the latter has led to some of the most important advances.

The earliest experiments were based on comparisons among those families and generations that most readily sprang from an initial cross between a pair of pure-breeding lines, for example, the selfing (F) series (Mather and Vines, 1952; Hayman, 1960). But it was soon realised that any design based solely upon inbreeding provided so little information on the dominance (H) and other non-additive components of variation that in most circumstances they could only be relied upon to provide repeatable estimates of the additive (D) and environmental (E) components (Mather and Jinks, 1971).

In the same way the earliest experiments adopted the natural group, the family, as the unit of randomisation and replication. But again it soon became apparent that this was introducing a complexity into the specification of the environmental components of variation and ambiguity into the interpretation that was unnecessary in species where complete randomisation of individuals was practicable (Jinks, 1954). It was also increasing the dimensions of the experiments required to give the same amount of information about the genetical components of means and variances.

It was only after conducting and analysing these early experiments (Mather and Vines, 1952; Jinks, 1954, 1956; Breese, 1956; Opsahl, 1956; Nelder, 1960) in which all the normal statistical requirements had been met, that these limitations on biometrical genetical interpretation and hence the need for improved experimental and breeding designs became apparent. Subsequent developments have underlined the considerable scope that existed, and still exists, for increasing the information obtainable on any source or sources of variation of special interest by choosing the most efficient material, breeding programme and experimental design for the purpose.

The biology of the material itself, for example, generation time, litter size, duration of essential maternal care, number of seed per pollination, number of flowers per plant, size and subsistence cost per individual $etc.$ all impose relatively inflexible restraints on the choice of breeding programme and experimental design. Detailed consideration of these factors is beyond the scope of this account. In general, however, an annual plant species which produces many flowers and sets seed equally well whether self- or cross-pollinated imposes almost no restraints while a large mammal with single births imposes the most.

The two simplest of the more widely used designs for a biometrical genetical analysis of a diploid species are the so-called basic generations if the material available is two or more pure-breeding lines and random biparental matings if it is a randomly mating population. The basic generations consist of the F_2 and first backcrosses, B_1 and B_2, of an F_1 cross between a pair of pure-breeding lines (P_1 and P_2). If all crosses are made reciprocally the means and variances of the P_1, P_2, F_1, F_2, B_1 and B_2 families will permit the detection of non-allelic interactions, genotype × environmental interactions, sex-linkage and maternal effects. They will also permit the estimation of all the components of the means for any source of variation which is significant but only the additive genetic and dominance components (D, H, and F) and the additive environmental component (E) of the variances and some of the contributions that sex-linkage, maternal effects, and genotype × environmental interactions may make to the variances (Mather and Jinks, 1971). For the analysis of means it is, therefore, a relatively powerful design. For the variances, however, the estimates will be biased in the presence of non-allelic interactions and genotype × environment interactions.

Random biparental matings produce a randomly mating population within which the individuals can be divided into full-sib families. The total variation can, therefore, be subdivided into the variation between full-sib family means (the full-sib covariance) and the mean variation within full-sib families. With only these two variances a maximum of two components can be estimated. By adding the parent-offspring covariance to the available statistics and using complete individual randomisation all three components of an additive genetic, dominance and additive environmental model of the variances and covariances (D_R, H_R, and E) can be estimated (see chapter by Mather in this volume; Kearsey, 1965; Mather and Jinks, 1971). Designs which improve on this situation by generating other degrees of relationship, such as half-sib, among the offspring of a randomly mating population include North Carolina designs 1 and 2 (Comstock and Robinson, 1952) and diallel sets of crosses (Dickinson and Jinks, 1956). The NC2 and diallel designs provide independent tests of significance of the additive genetic and dominance components of variation and all three designs provide estimates of these components and of the environmental components of variation. However, in the progression from random biparental matings to diallels the demands imposed by the breeding design become so exacting that diallels are restricted to the most flexible species, as defined earlier. But even with the most sophisticated of these designs the analysis of a randomly mating population has to proceed on the assumption that an additive, dominance model of the heritable variation is adequate because these are no ways of testing for or accommodating in the analysis additional sources of variation other than possibly sex-linkage and maternal effects (Mather and Jinks, 1971). Furthermore, while the estimates of the random mating form of the additive and dominance components (D_R and H_R) have some predictive value (section VIIIB) they are not suitable for example for estimating the dominance ratio or for predicting the properties of inbred derivatives of the population. One would never choose, therefore, to work with randomly mating populations if the main purpose was to investigate the nature of the action, interaction, linkage relationships and reaction to environmental variability of the genes controlling quantitative variation or to make specific rather than general predictions about the consequences of selection, inbreeding and backcrossing.

There are, of course, other ways of investigating randomly mating populations which involve either inbreeding a random sample (Mather and Jinks, 1971; Kearsey, 1970; Hillel *et al*. 1972), or making test-crosses to pure-breeding lines or their F_1's (Comstock and Robinson, 1952; Kearsey and Jinks, 1968). In most circumstances these designs permit more powerful analyses than their random mating equivalents but the test-cross in particular is more restricted in its applications because of the need for pure-breeding testers of the right kind. But with the exception of the triple-test cross (Kearsey and Jinks, 1968) these designs provide no test of the assumption that all the variation can be attributed to additive and dominance gene action and additive environmental effects.

While the multiple mating designs were developed primarily to strengthen the analysis of randomly mating populations, and it is in this context that they have been most widely used and discussed so far, they are not only applicable to other kinds of population for example, inbred and F_2 populations, but they are more effective in such applications.

Broadly speaking there are four categories of populations that the multiple mating design might be used to analyse:

1. Randomly mating populations
2. F_2 populations derived from a cross between two pure-breeding lines (or a randomly mating population with equal gene frequencies
3. Pure-breeding lines with arbitrary gene frequencies as might be obtained by inbreeding a randomly mating population
4. Pure-breeding lines with equal gene frequencies as might be obtained by inbreeding an F_2 of a cross between two pure-breeding lines (or a randomly mating population with equal gene frequencies).

With populations of types 3 and 4 the heritable components make a two-fold greater contribution to the comparison between families that are used for their detection and estimation than do populations of types 1 and 2. Furthermore, with populations of types 3 and 4 comparisons within families or between replicate families, which are the sources of error mean squares, contain no heritable components and hence are smaller than those with populations of types 1 and 2 which do. Irrespective therefore of which of the various multiple mating and test cross designs are used it is much easier to detect and obtain reliable estimates of the heritable components of variation in populations of types 3 and 4 than in populations of types 1 and 2. Over and above this, however, the contribution that each of the loci make to the heritable components of variation is at its greatest where, as in populations of types 2 and 4, there are equal gene frequencies. But more importantly, where these populations have been derived from an initial cross between a pair of pure-breeding lines, the latter and their F_1 cross are the ideal testers in the most sophisticated of all these designs, the triple-test cross (Kearsey and Jinks, 1968; Jinks and Perkins, 1970; Pooni and Jinks, 1976; Pooni et al.,

1978). Not surprisingly, therefore, the optimal, practical designs for most purposes are where the advantages of using material with equal gene frequencies can be combined with those of the best multiple mating and test-cross breeding programme, for example, diallel sets of crosses between purebreeding lines (Jinks and Hayman, 1953; Mather and Jinks, 1971, 1977) and triple-test crosses on F_2 populations (Jinks and Perkins, 1970).

An alternative approach was initiated by building upon the basic generations by further rounds of selfing, backcrossing and sib-mating among the P_1, P_2, F_1, F_2, B_1, and B_2 families (Mather and Vines, 1952; Opsahl, 1956; Hayman, 1960; Hill, 1966). This has culminated in a design consisting of the 21 generations produced by all possible matings within and between these six families. In addition to the six families themselves, these 21 generations include the three triple test crosses on the F_2, B_1 and B_2 families produced by crossing samples of each to P_1, P_2 and F_1, randomly mated and selfed F_2's, selfed first backcrosses and second reciprocal backcrosses (B_{11}, B_{12}, B_{21} and B_{22}) (Jinks and Perkins, 1969; Perkins and Jinks, 1970). This design permits the detection and estimation of the magnitude and direction of all the sources of variation identified in Section II both at the level of means and of variances. In practice, analysis of the means has been used to distinguish between trigenic interactions and linkage of genes displaying digenic interactions only, while analysis of the variances has been used to detect non-allelic interactions, to measure the magnitude and phase of linkage and the relative sensitivities of different kinds of gene action to micro-environmental variation. Because many of the advantages of this design result from combining the basic generations with the F_2 triple-test cross this combination has become the standard design for initiating a biometrical genetical analysis in a new situation.

In parallel with these developments attention has been given to the more quantitative aspects of the design of experiments, for example, the minimum size of experiment required to achieve a particular objective and the optimal distribution of resources among the different families and relationships that constitute a single experiment within a fixed total. Where the objective is specific and restricted to a few properties, *e.g.* the detection of dominance or non-allelic interactions, the measurement of heritability or the dominance ratio, *etc.*, the precise dimension and composition of the data which should be collected have been established by a combination of theoretical considerations and computer simulations often supported by the analysis of experimental data (Kearsey, 1970; Eaves, 1970, 1972; Pederson, 1971; Pooni and Jinks,

1976). The same approach has been used to determine the relative efficiencies of a number of different experimental designs by comparing the minimum dimensions required to achieve the same objectives (Kearsey, 1970; Mather and Jinks, 1971).

Another approach has been to define as the objective the generation of as much information as possible from an experiment of fixed total size and composition by varying the relative frequencies of the various kinds of families and family sizes (Jinks and Perkins, 1969). These developments are still in their infancy but from the improvements in the quality of data already achieved there can be no doubt of the worthwhileness of considering these aspects of experimental design through theoretical models and computer simulations before embarking on an experimental breeding programme. For these purposes the simulation package developed by Dr. M.J. Kearsey has proved to be invaluable and examples of its use in the design of experiments have been given by Pooni, Jinks, and Cornish (1977), Eaves, Last, Martin, and Jinks (1977), and Martin, Eaves, Kearsey, and Davies (1978).

VIII. APPLICATIONS

The procedures of biometrical genetics can be used to achieve two broad objectives. The first is to provide detailed information on the genetical systems which control quantitative or continuous variation, increasingly referred to as the genetical architecture, to obtain a better understanding of how such systems might have evolved, how the variation is maintained and how it might respond to selection. The second is to use the information gained from the early stages of a breeding programme to interpret how any observed changes in the mean, variance, skewness and kurtosis from one generation to another have come about and on the basis of this interpretation to predict what further changes might be expected from any given breeding programme or selection regime that might be subsequently imposed.

A. *Genetical Architecture*

A biometrical genetical investigation is the only practical means of determining the genetical architecture of a continuously varying character outside of a few well known species. That different types of natural selection should lead to characteristically different genetical architectures has been a developing theme of Mather and his school since 1943 (*e.g.*,

Mather, 1943, 1953, 1960, 1973; Breese and Mather, 1960; Kearsey and Kojima, 1967; Broadhurst and Jinks, 1974; see also Mather in this volume). Data supporting the principal conclusions has been assembled over the intervening years (see Mather, 1973; Kearsey and Kojima, 1967; and Broadhurst and Jinks, 1974), and there are now many examples of the past history of selection and future response to selection being inferred from a detailed analysis of the genetical architecture using the techniques of biometrical genetics.

Information on any aspect of the genetical architecture is, of course, helpful in drawing these inferences but the following properties have been found repeatedly to be particularly relevant.

1. Dominance properties. The presence or absence of dominance ($H > 0$ or $= 0$) and if present, whether it is directional or ambidirectional ($|[h]| > 0$ or $= 0$) are important considerations in deciding between a past history of directional or stabilising selection.

2. Type of non-allelic interaction. Whether the non-allelic interaction, if present (*i.e.* $[i]$, $[j]$ and $[l]$ significant), is of a duplicate type ($[h]$ and $[l]$ of opposite sign), is particularly important in recognising characters which have had a long history of directional selection for an extreme expression and hence are components of fitness.

3. Linkage disequilibrium. Whether linkage, if present (the genetic components of variation change with rank), is due to an excess of genes in the repulsion or dispersed phase (the additive genetical component of variation increases with rank) can be indicative of a past history of stabilising selection.

4. Number and nature of effective factors. The number and nature of the genes or more strictly of the effective factors (Mather, 1949; Mather and Jinks, 1971, 1977) that control continuous variation has recently received both critical and constructive attention. Biometrical genetics offers a number of ways and the only generally applicable ways of estimating their number by comparison between first and second and second and fourth degree genetical components. Recently it has offered a new procedure, genotype assay (Jinks and Towey, 1976; Towey and Jinks, 1978). The estimates obtained when combined with the results of related investigations of fungal species have thrown considerable light on the nature of the genes and the composition of effective factors both of which are discussed in more detail in the chapters by Mather and by Caten in this volume.

B. *Prediction*

The expected means, variances, covariances, skewness, *etc.* of every possible family, generation and population can be expressed in terms of the heritable and non-heritable components of these statistics. Following the estimation of these components the expected magnitude of each statistic can be calculated not only for the families, generations and populations that have been observed up to that time and which may themselves have contributed to the estimates but also for any family, generation or population that could conceivably be derived from them in the future by any controlled breeding programme one might wish to impose. The applications of biometrical genetics in prediction are all special cases of this general property which in turn is a consequence of the way in which the components have been defined (section II). Indeed, as we have already noted (section I) it is because the heritable components have been defined in terms of genetical rather than solely on statistical principles that they have this value in interpretation and prediction. The following are a few illustrations. Others appear in the chapters by Mather and by Caten.

1. *Genotype × Environment Interaction.* This is in some ways an atypical illustration in that non-heritable as well as heritable components are involved in the predictions, but it brings out the contrast between genetically and statistically defined components. By partitioning the genotype × environment interaction contributions to means and variances according to the source of heritable variation that is involved in the interaction, changes in the cause and magnitude of the interaction from one generation to another can be predicted (Buci Alanis *et al.*, 1969; Perkins and Jinks, 1970). Because, however, the environmental components involved in these interactions are defined solely on statistical considerations, there is no corresponding ability to predict from one environment to another. Nevertheless there are biometrical genetical procedures for making simultaneous predictions over generations and environments whenever a simple relationship can be empirically found between the genotype × interaction component for each source and either the magnitude of the additive environmental component or some physical environmental variable over the range of environments studied (Bucio Alanis, 1966; Bucio Alanis *et al.*, 1969; Fripp, 1972; Perkins, 1972; Jinks and Connolly, 1975). Such relationships have been found in practice and successfully used to predict the performance of genotypes in environments which are related to, but different from, those in which they had been previously observed.

2. *Heterosis*. A more typical illustration is provided by biometrical genetical analyses of heterosis or hybrid vigour, defined as the superiority of an F_1 over its better parent (P_1 or P_2 as appropriate). This has an expectation which can be expressed in terms of the components of means, *e.g.* $[d]$, $[h]$, $[i]$, $[l]$, *etc.* and its environmental dependence specified in terms of the genotype × environmental components of means, *e.g.* g_d, g_h, g_i, g_l, *etc.* (Jinks and Jones, 1958; Bucio Alanis *et al.*, 1969; Mather and Jinks, 1971). Providing that there are no significant trigenic or higher order interactions all of these components can be estimated from the basic generations raised in the environments of interest. To test the adequacy of this model it would in general be necessary to add further generations such as second backcrosses or F_3 (Jinks and Perkins, 1969; and section VII).

It is, therefore, a relatively straightforward matter to determine the relative contribution that each source of variation is making to the observed heterosis and hence to predict how it will respond to, say, recurrent backcrossing or inbreeding. Where it can be empirically demonstrated that the genotype × environment interaction components are simply related to a biological measure of the environment, the additive environmental component, or to some physical attribute of the environment, it is possible to predict how the magnitude of the heterosis will respond to an improvement or a deterioration in the environment.

However, to establish whether dispersion of dominant alleles or overdominance is the underlying cause of the heterosis, and hence the prospects of obtaining inbred lines which are superior to the F_1, it is necessary to examine the components of variance. And since the dominance ratio plays a key role in this decision, it is important that the dominance component be estimated with the same reliability as the additive genetic component by using a design such as an F_2 triple test cross. It is also important to establish whether the estimate is likely to be biased upwards or downwards by epistasis or linkage disequilibrium. Again, an F_2 triple test cross can provide tests of both. By combining the information from the basic generations and the triple test cross, the probability of obtaining superior inbreds and the limits to their performance can be quantified (section VIII B3).

In general, biometrical genetical analyses of heterosis have drawn attention to the important contribution of non-allelic interaction and the absence of clear evidence of overdominance as opposed to dispersion as a cause of heterosis. The highest levels of heterosis are almost invariably associated with the greatest involvement of non-allelic interactions (*e.g.*, Jinks and Jones, 1958).

In the few instances where the environmental dependence of heterosis has been investigated, the additive and dominance gene actions have shown different linear relationships with the environmental variation. As a result the magnitude of the heterosis has changed in a predictable way with the changes in the environment and the kinds of environment in which heterosis would be at its greatest and least correctly identified (Bucio Alanis *et al.*, 1969).

3. *The Properties of Recombinant Inbred Lines.* The expected distribution of the recombinant inbred lines that can be derived by inbreeding without selection from the F_2 of a cross between a pair of pure-breeding lines can be given in terms of the components of means, variances, skewness, *etc.*, which can be estimated during the early stages of the breeding programme (Jinks and Perkins, 1972; Jinks and Pooni, 1976; Pooni *et al.*, 1977; Pooni and Jinks, 1978). In general, however, interest will be centred on a selected sample rather than on the whole distribution. For example, if the F_1 is superior to both parents the main interest may be in what proportion, if any, of the inbreds will be superior to the F_1; if on the other hand there is no heterosis the main interest may be in what proportion, if any, of the inbreds will be superior to the better of the pure-breeding parents.

If an additive, dominance model of the heritable components is adequate all that is required for making these predictions is estimates of m, [d], [h] and $\sqrt{D}$ for each character or set of independent characters. For each pair of correlated characters an estimate of the additive genetic covariance (D_{12}) is required in addition. All of these are readily obtainable from an F_2 triple test cross or the basic generations. The worthwhileness of proceeding with a breeding programme designed to produce superior inbreds can, therefore, be judged on the basis of these predictions at an early stage as can also the relative merits of two or more different crosses as sources of superior inbreds.

Non-allelic interactions, genotype × environment interactions and autosomal linkage could complicate these simple, and in practice, reliable predictions. Of these, however, only non-allelic interaction has led to unsatisfactory predictions if no allowance is made for its presence (Jinks and Pooni, 1976; Pooni *et al.*, 1977). And when allowance has been made for its contribution to the mean by including [i] and [l] in the expectations, all the predictions have been satisfactory.

Corresponding predictions can be made about the distribution of the pure-breeding lines that can be derived by inbreeding from a randomly mating population provided that the estimate of the additive genetic component of variation is that obtained from a triple test cross. The random mating form of the additive component (D_R) which is the one obtained from the other mating systems (section VII) is not suitable for making these predictions unless it is known that the gene frequencies are equal at all loci or that there is no dominance.

4. *Response to Selection*. Narrow heritability, that is the proportion of the total variation in a population that is fixable, which in a simple situation is the additive component of the variation ($\frac{1}{2} D_R$ for a randomly mating population, $\frac{1}{2} D$ if the population has equal gene frequencies) can be used to predict the short term response to selection (Falconer, 1960; see also other chapters in this volume). The narrow heritability can be measured directly on a population from the regression of offspring family mean on parental mean. It can also be estimated indirectly from other relationships such as the intraclass correlation between half-sib family means. Unless, however, an experimental design is used which permits a complete partitioning of the variation into all of its significant components, it will not be possible to distinguish between different causes of, for example, a low narrow heritability which could arise either from a high dominance or non-allelic interaction component at one extreme or from a high non-heritable component with no non-additive heritable component at the other.

With estimates of the various heritable and non-heritable sources of variation the narrow heritability can be estimated not only for the initial population, but for any population that can be derived from it by a controlled breeding programme. One can, therefore, begin to assess the desirability of modifying the initial population, for example, by inbreeding, before embarking on a selection programme.

Reliably estimated, the narrow heritability has considerable utility as a simple practical guide. No single ratio, however, can hope to give more than a superficial summary of the total information that a biometrical genetical analysis can yield that has a direct bearing on the capacity of the genetical system to respond to selection and its limits.

IX. CONCLUSIONS

In this brief account of the philosophy, procedures, applications and achievements of biometrical genetics I have set out to show -- in a way that I believe a general geneticist can follow -- that there are no limits to the complexity of the control of quantitative variation that it can elucidate. Theoretical expectations are available which cover almost every conceivable situation and experimental designs and associated statistical analyses are available for detecting and often estimating the contributions of all known sources of variation.

Applications of biometrical genetics have shown that the underlying control of quantitative variation is frequently complex. Analytical approaches that cannot accommodate these complexities are, therefore, on the one hand unrealistic and on the other potentially misleading. There are, of course, many applied situations where the magnitude of the biases introduced by ignoring the more complex interactions is insufficient to have any practical consequences. But this can only be established empirically in each new situation by having the means of detecting these biases and obtaining some estimate of their order of magnitude. Biometrical genetics alone provides these means in the widest range of situations.

REFERENCES

Breese, E.L. (1956). *Heredity 10*, 323-343.
Breese, E.L., and Mather, K. (1960). *Heredity 14*, 375-400.
Broadhurst, P.L., and Jinks, J.L. (1974). *In* "The Genetics of Behaviour" (J.H.F. Van Abeelen, ed.), pp. 43-63. North Holland, American Elsevier, New York.
Bucio Alanis, L. (1966). *Heredity 21*, 387-397.
Bucio Alanis, L., Perkins, J.M., and Jinks, J.L. (1969). *Heredity 24*, 115-127.
Cavalli, L.L. (1952). *In* "Quantitative Inheritance" (E.C.R. Reeve and C.H. Waddington, eds.) pp. 135-144. H.M.S.O., London.
Comstock, R.E., and Robinson, H.F. (1952). *In* "Heterosis" (J.W. Gowen, ed.) pp. 494-516. Iowa State College Press, Ames, Iowa.
Dickinson, A.G., and Jinks, J.L. (1956). *Genetics 41*, 65-77.
Eaves, L.J. (1970). *Brit. J. Math. Statist. Psychol. 22*, 131-147.
Eaves, L.J. (1972). *Heredity 36*, 205-214.

Eaves, L.J. (1976). *Heredity 36*, 205-214.
Eaves, L.J., Last, K.A., Martin, N.G., and Jinks, J.L. (1977). *Brit. J. Math. Statist. Psychol. 30*, 1-42.
Falconer, D.S. (1960). "Introduction to Quantitative Genetics." Oliver and Boyd, Edinburgh.
Fripp, Y.J. (1972). *Heredity 28*, 223-238.
Gale, J.S., Mather, K., and Jinks, J.L. (1977). *Heredity 38*, 47-51.
Gilbert, N.E. (1967). *Biometrics 23*, 45-49.
Griffing, B. (1956). *Australian J. Biol. Sci. 9*, 463-493.
Hayman, B.I. (1960). *Biometrics 16*, 369-381.
Hayman, B.I., and Mather, K. (1955). *Biometrics 10*, 69-82.
Hill, J. (1966). *Heredity 21*, 85-120.
Hillel, J., Simchen, G., and Jinks, J.L. (1972). *Theoret. Pop. Biol. 4*, 466-490.
Jayasekara, N.E.M., and Jinks, J.L. (1976). *Heredity 36*, 31-40.
Jinks, J.L. (1954). *Genetics 39*, 767-788.
Jinks, J.L. (1956). *Heredity 10*, 1-30.
Jinks, J.L. (1978). *Heredity 40*, 171-173.
Jinks, J.L., and Connolly, V. (1975). *Heredity 34*, 401-406.
Jinks, J.L., and Hayman, B.I. (1953). *Maize Cooperation Newsletter 27*, 48-54.
Jinks, J.L., and Jones, R.M. (1958). *Genetics 43*, 223-234.
Jinks, J.L., and Perkins, J.M. (1969). *Heredity 24*, 465-475.
Jinks, J.L., and Perkins, J.M. (1970). *Heredity 25*, 419-429.
Jinks, J.L., and Perkins, J.M. (1972). *Heredity 28*, 399-403.
Jinks, J.L., and Pooni, H.S. (1976). *Heredity 36*, 253-266.
Jinks, J.L., and Towey, P.M. (1976). *Heredity 37*, 69-81.
Jinks, J.L., Perkins, J.M., and Gregory, S.R. (1972). *Heredity 28*, 363-377.
Kearsey, M.J. (1965). *Heredity 20*, 205-235.
Kearsey, M.J. (1970). *Heredity 25*, 529-542.
Kearsey, M.J., and Jinks, J.L. (1968). *Heredity 23*, 403-409.
Kearsey, M.J., and Kojima, K. (1967). *Genetics 56*, 24-37.
Killick, R.J. (1971). *Heredity 27*, 175-188.
Martin, N.G., Eaves, L.J., Kearsey, M.J., and Davies, P. (1978). *Heredity 40*, 97-116.
Mather, K. (1941). *J. Genet. 41*, 159-193.
Mather, K. (1943). *Biol. Rev. 18*, 32-64.
Mather, K. (1949). "Biometrical Genetics." Methuen, London.
Mather, K. (1953). *Symp. Soc. Exp. Biol. 7*, 66-95.
Mather, K. (1960). "International Colloquium on Evolution and Genetics." pp. 131-152. Acad. Naz. dei Lincei, Rome.
Mather, K. (1973). "Genetical Structure of Populations." Chapman and Hall, London.

Mather, K., and Jinks, J.L. (1963). *Nature 198*, 314-315.
Mather, K., and Jinks, J.L. (1971). "Biometrical Genetics", 2nd edition. Chapman and Hall, London.
Mather, K., and Jinks, J.L. (1977). "Introduction to Biometrical Genetics." Chapman and Hall, London.
Mather, K., and Jones, R.M. (1958). *Biometrics 14*, 343-359.
Mather, K., and Vines, A. (1952). *In* "Quantitative Inheritance" (E.C.R. Reeve and C.H. Waddington, eds.), pp. 49-80. H.M.S.O., London.
Nelder, J.A. (1960). *In* "Biometrical Genetics" (O. Kempthorne, ed.), pp. 139-158. Pergamon Press, London.
Opsahl, B. (1956). *Biometrics 12*, 415-432.
Pederson, D.G. (1971). *Heredity 27*, 247-264.
Perkins, J.M. (1972). *Heredity 29*, 51-70.
Perkins, J.M., and Jinks, J.L. (1970). *Heredity 25*, 157-177.
Perkins, J.M., and Jinks, J.L. (1971). *Heredity 26*, 203-209.
Pooni, H.S., and Jinks, J.L. (1976). *Heredity 36*, 215-227.
Pooni, H.S., and Jinks, J.L. (1978). *Heredity 40*, 349-361.
Pooni, H.S., Jinks, J.L., and Cornish, M.A. (1977). *Heredity 38*, 329-338.
Pooni, H.S., Jinks, J.L., and Jayasekara, N.E.M. (1978). *Heredity 41*, 83-92.
Towey, P.M., and Jinks, J.L. (1978). *Heredity 39*, 399-410.
Van Der Veen, J.H. (1959). *Genetica 30*, 201-232.

The Uses and Limitations of Selection

AN OVERVIEW OF SELECTION THEORY AND ANALYSIS

Clifford Johnson

Department of Zoology
University of Florida
Gainesville, Florida

I. INTRODUCTION

The action of selection on genetic systems is one of the richest areas of concept in all biology. The theory and its documentation can be traced back to the early controversies surrounding evolution, and its expansion and sophistication continues to the present time. In this chapter some of the major features of selection theory are introduced and described without technical, rigorous proof or review of experimental confirmation. A large number of investigators have made important contributions over a long period, and reviews of the extensive literature are cited at the end of each subsection.

II. SELECTION AND VARIABILITY

Selection on polygenic variation changes gene frequencies or maintains gene combinations in non-random associations. Effects of these processes are judged from visible variation, and the underlying genetic system controlling this variation has overriding importance when applying or interpreting selection. Phenotypic variance, σ_P^2, is conventionally viewed as the sum of three components, variances due to environmental action, gene-environment interaction, and genetic segregation, σ_E^2, σ_{GE}^2, and σ_G^2, respectively. The latter component, our main concern, is further partitioned into variances arising from additive, dominant and epistatic, non-allelic gene actions, σ_A^2, σ_D^2 and σ_I^2, respectively.

Selection effects, in a given generation, are limited by the expressed or free variation of the underlying genetic system. The genetic system also accommodates hidden variation potentially available in later generations. This potential results from additive balance per locus (heterozygosity) and additive balance between loci (balanced homozygotes). If we let additivity exist for two loci such that AABB = +2, AaBB = +1, AaBb, AAbb, and aaBB = 0, Aabb and aaBb = -1, and aabb = -2, then a population consisting of equal numbers of AABB and aabb individuals has the maximum σ_P^2 of 4.0. If an F_1 generation is obtained from only AABB × aabb crosses, its σ_P^2 is zero and the σ_P^2 of the subsequent F_2 is only 1.0, or ¼ of the original. The remaining 3/4 is not lost, but exists as hidden variation consisting of ½ heterozygotic and ¼ homozygotic potential variation. If the F_2 (where all gene frequencies are equal) mate randomly, this level of variability persists. However, if continued inbreeding follows, σ_P^2 increases to 2.0 while ½ of the original value has become locked into the homozygotic potential forms, AAbb and aaBB, that were originally absent.

An increase in the number of loci modifies this property into the ratio: 1 free to n heterozygotic potential to n-1 homozygotic potential states, where n is the number of loci. This ratio applies both with and without equal gene frequencies. Thus, just ten loci give a σ_P^2 expressing only 1/20 of the possible variability; with 20 loci, the σ_P^2 shows only 1/40 of the possible variation. In polygenic systems based on additivity, a large total variation may, therefore, exist while exposing only a small amount as free and sensitive to selection in any one generation. A high level of heterozygosity, thus, does not give an increasing exposure to selection. This conclusion holds even without further restraining the release of free variation by way of linkage.

If factors of a polygenic complex are linked, the flow of variation between free and potential states may be significantly affected. The flow of potential variation into the free state arises directly from segregation and recombination within heterozygous genotypes. Where loci have alleles giving high (+) and low (-) phenotypic effects, and where such alleles are linked in repulsion, the formation of homozygotes contributing to free variability is largely dependent on the level of recombination. Theory predicts that a moderate release of free variation and a large store of potential variability is most favorable, and this system clearly favors the evolution of repulsion linkage. Indeed, we might expect the alleles along a given chromosome to alternate in a - + - + pattern while the homolog would have the complementary + - + - pattern. This alternating sequence will require multiple crossovers to build gametes capable of forming homozygotes, thus further retarding the rate of release of free variation.

If a large population has two independent loci having alleles A, a, B, and b, with gene frequencies p, q, r, and s, respectively, and if the individuals mate at random with respect to these alleles, the gametic frequencies will be pr, ps, qr, and qs, for AB, Ab, aB and ab, respectively. These values remain stable if no disturbing force, such as selection, affects the population. Now let a second population have similar gene frequencies, assumed equal for our example, but linked in repulsion, Ab/aB, for all members. Recombination will occur at a rate m. For the first generation, the non-recombinant gametes, Ab and aB, each have a frequency of $\frac{1}{2}(1-m)$, while the recombinant gametes, AB and ab, each occur at a frequency of $\frac{1}{2}m$. Consequently, frequencies of each gamete change in the following generations and approach the same values as occur in the population with no linkage. Even with tight linkage, the two populations would soon have gamete frequencies determined solely by gene frequencies.

If selection intervenes favoring non-recombinants, however, the recombinants may be lost at a rate similar to their formation and, thus, preserve the non-random gamete frequencies. Gametic disequilibrium enhanced by linkage is termed linkage disequilibrium and is one potential means of detecting selection in natural populations.

Documentation for these properties of variability can be found in Falconer (1960), Mather and Jinks (1971), Mather (1973), Li (1976), and Spiess (1977).

III. SELECTION PATTERNS

Selection typically favors one or a small number of classes conventionally identified as the optimal phenotypes. Selection patterns vary in relation to the position of the optimal class and the mean of expressed variation. Stabilizing selection operates when the optimal class lies close to the mean and variability is reduced symmetrically around it. Thus, parents are favored that produce phenotypically average offspring with least variation. Directional selection acts upon an optimum distinctly different from the original optimum and shifts the mean towards that value. Selection may also recognize two or more optimal classes within one distribution.

This produces a bimodal or multimodal distribution following selection. Such selection is termed disruptive or diversifying selection. The effects and applications of these patterns on genetic variability are discussed briefly below.

IV. DIRECTIONAL SELECTION

Most traits in natural populations, other than viability and fertility, probably exist under stabilizing selection, though changes in environmental conditions or heterogenous environments lead to directional selection. The analysis of directional selection is based upon the parameters summarized in Figure 1.

From a parental population having a mean of $\overline{Y}_p$, let all individuals having a phenotypic value of x or greater be selected as parents for the next generation. The mean of

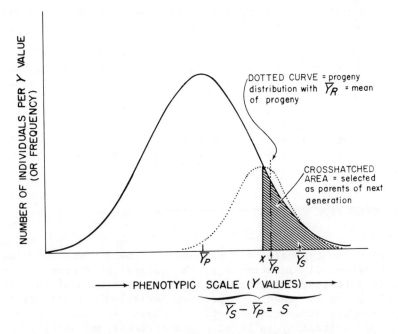

FIGURE 1. *A population having a normally-distributed quantitative trait, Y, with a mean of $\overline{Y}_p$. Individuals at point x or greater and with mean $\overline{Y}_S$ are selected as parents. The dotted curve gives their offspring distribution for the trait with mean $\overline{Y}_R$. The selection differential is S.*

selected parents is $\overline{Y}_S$ and that of their progeny, typically less, is $\overline{Y}_R$. Then,

$$\overline{Y}_S - \overline{Y}_P = S,$$

where S is the selection differential, and

$$\overline{Y}_R - \overline{Y}_P = R,$$

where R is the selection response.

If R/S = 1.0, then all phenotypic variation appears to be due to additive genetic variation. However, if R/S < 1.0, the ratio gives the mean degree to which additive genetic variation controls the phenotype; this estimate is the realized heritability, h^2. The principal value of h^2 consists in its predictive property for R, since heritability gives the expected similarity between relatives and, thus, the available additivity for selection to act upon. Heritability in this sense is σ_A^2/σ_P^2.

In the definitions given above, S and R were given in phenotypic scale units and, thus, may vary according to the proportion of the population selected as parents. These parameters are standardized when given in units of the phenotypic standard deviation, S/σ_P and R/σ_P. From the modified expression we find $R/\sigma_P = h^2(S/\sigma_P)$. The term S/σ_P is the intensity of selection, i, giving $R/\sigma_P = ih^2$. The measure of selection, i, is important in applied problems and its following property should be noted. When using the same phenotypic scale unit, a more variable population has a smaller σ_P than does a less variable population. The value of i, therefore, depends on the proportion of the total selected as parents. In this sense, i is given by height of the ordinate at the point of truncation, x, in Figure 1, divided by the proportion selected. Note that i does not have units between zero and one. The $R/i\sigma_P$ expression provides a usable estimate of h^2; however, such values, by more analysis, are seen to be potentially compromised by dominant gene action. Ideally, a direct measure of σ_A^2 is required, and biometrical methods provide means for finding this value (Mather and Jinks, 1971).

Several properties of h^2 deserve further consideration. First, h^2 is a population parameter and generalizations cannot be applied to specific individuals. In addition, the value of h^2 is not constant. Theoretically, h^2 declines as selection accumulates homozygotes and variation becomes fixed. The response predicted by h^2 is, in theory, valid for only one generation, though in practice its predictive value can hold for several generations. Finally, correlated and accelerated responses are important processes in directional selection and h^2 gives no prediction of their likelihood or nature.

From this brief review, the efficiency of selection is seen to involve available free variation and its rate of release from the hidden potential, the rigour of selection where higher S and i values result from smaller proportions selected as parents, the number of effective factors, and non-heritable effects. Design of a selection program and choice of breeding stocks can often take advantage of these features. A higher initial response is often possible if several different stocks are initially available. The stock having the highest initial frequency offers more free variation. Response will also be greater if the trait's underlying polygenic system is independent of other traits that may otherwise give a correlated response in lowering fitness. Polygenic systems clearly do not exist as independent components of the full genetic makeup; however, several possible phenotypes may achieve a given economic objective but show quite different responses. As an example, yield in corn could be increased by larger ears or larger plants with multiple, smaller ears, yet selection response is unlikely to be similar for these two paths to one objective. Phrased in the above terms, a higher response is expected by increasing the correlation between breeding value and selection criteria. Practical insight on breeding values involves pedigree history and progeny testing, and it may often require considerable time to accumulate.

The technical treatment of directional selection is discussed in Lerner (1958), Falconer (1960), Pirchner (1969), and Spiess (1977).

V. CORRELATED RESPONSES AND SELECTION LIMIT

Directional selection often encounters response in traits not intentionally selected. The most typical correlated responses consist of reduction in fertility, viability, and other fitness characters. Genetically, this phenomenon may involve linkage between the polygenes under selection and polygenes affecting other traits. Little basis exists, therefore, for predicting the nature or onset of correlated responses.

If linkage is the basis of a correlated response lowering fitness, recombination within the gene complex may lead to genotypes where further response is possible. The problem consists in maintaining a stock at a given level of response, while fitness is low, over a period sufficient for the necessary recombination.

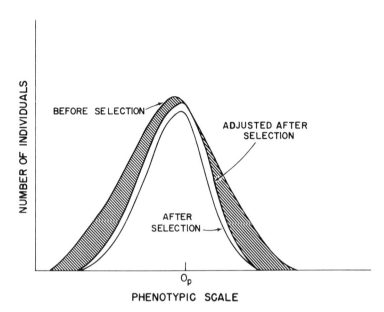

FIGURE 2. *A population having a normally-distributed quantitative trait where the optimal phenotype, O_p, occurs close to the mean value prior to selection. The adjusted after-selection curve depicts O_p with maximal survival and the shaded area is the proportionate loss to selection.*

Correlated responses are discussed in the previously cited references on directional selection. In addition, a very good treatment is presented in Lee and Parsons (1968).

VI. STABILIZING SELECTION

Stabilizing selection operates when the majority of the population closely approximates the optimal phenotype. The mode of operation and rationale for its analysis can be explained using Figure 2. The population, prior to selection, is characterized by the distribution of the uppermost curve and by the lowermost curve for after-selection properties. The area between these curves represents the loss by inviability or infertility due (1) to ecological control of population size and (2) to differential effects between genotypes, *i.e.* to selection. Thus, the principal result of stabilizing selection is a decrease in phenotypic variance.

The optimal phenotype is that value on the horizontal scale having least total loss; it is shown sligtly displaced from the mean as O_p in Figure 2. If each point on the after-selection curve is proportionately increased by the amount needed to place O_p on the before-selection curve, the adjusted after-selection curve gives the distribution reflecting only loss by selection. The percentage loss by selection appears in the literature as the phenotypic load or as the intensity of selection, I. A number of methods for estimating I are discussed in Cook (1971), Johnson (1976), and Spiess (1977). Mather (1973) presents a general discussion of the effects of stabilizing selection.

VII. DISRUPTIVE SELECTION

The recognition of two or more optimal phenotypes within a single interbreeding population constitutes one of the most interesting forms of selection. The presence of two or more optima in one distribution leads to an increase in variance and bi- or multi-modal distribution. Experimental work on disruptive selection has largely used continuous variation such as bristle number in *Drosophila*. However, many situations concern discrete optimal classes without intermediate types. The development of an individual is, thus, channeled into one optimal phenotype or another by a developmental switch mechanism.

A switch system may involve environmental cues or be due to the genetic architecture of the trait. Where the optima alternate in time, a common genotype may characterize a population over several generations, but produce distinct phenotypes under different environmental conditions; seasonal polymorphism is a well-known phenomenon.

When optima co-exist, their relative frequencies may be critical to the function of each. Environmental cues may also operate in a few systems of this nature. Examples include nutritional control of caste in honeybees and environmentally controlled sex determination in some marine polychaete worms. The organization of a beehive provides a steady flow of information concerning the need of reproductives, workers, and other castes. These needs are reflected in diets given the developing young. In the polychaete worm, *Bonellia*, a young worm settling on the substrate close to an established female develops into a male, otherwise it develops into a female. Clearly, the frequency of young developing into males is higher when the existing frequency of females is high.

A switch built within the genetic system has the capacity

to adjust more finely to given frequencies. As an example, consider the classical case of sex determination involving one homogametic sex and one heterogametic sex. Sexes occur in equal frequencies, providing a clear functional dependency between the optima.

Mather (1973) identifies the problems and results of disruptive selection, while Thoday (1967, 1972) gives an excellent review summarizing both earlier and more recent work.

REFERENCES

Cook, L.M. (1971). "Coefficients of Natural Selection." Hutchinson and Co., London.
Falconer, D.S. (1960). "Introduction to Quantitative Genetics." Ronald Press, New York.
Johnson, C. (1976). "Introduction to Natural Selection." University Park Press, Baltimore.
Lee, B.T.O., and Parsons, P.A. (1968). *Biol. Rev. 43*, 139-174.
Lerner, I.M. (1958). "The Genetic Basis of Selection." Wiley, New York.
Li, C.C. (1976). "First Course in Population Genetics." Boxwood Press, Pacific Grove.
Mather, K. (1973). "Genetical Structure of Populations." Chapman and Hall, London.
Mather, K., and Jinks, J.L. (1971). "Biometrical Genetics." Cornell University Press, Ithaca.
Pirchner, F. (1969). "Population Genetics in Animal Breeding." Freeman, San Francisco.
Spiess, E.B. (1977). "Genes in Populations." Wiley, New York.
Thoday, J.M. (1967). *Genet. Res. 9*, 119-120.
Thoday, J.M. (1972). *Proc. Roy. Soc., B, 182*, 109-143.

COMPUTER SIMULATIONS

H. Geldermann and H. Gundel

Institut für Tierzucht
und Vererbungsforschung
Tierärztliche Hochschule
Hannover, Germany

I. INTRODUCTION

Genetic research on selection acting upon quantitative characters requires very complex considerations. Therefore, mathematical models necessitate simplifications and artificial assumptions which reduce the validity of the models when applied to natural situations. On the other hand, real experiments must also be limited to some extent, and in a complex field they only yield results concerning particular aspects. For this reason there is a void in quantitative genetics between theoretical reflections and experimentally verified facts. It is conceivable that this gap can be made smaller by the performance of simulations. Simulation can be defined as a modelling of real facts. In quantitative genetics, simulations are generally performed on the basis of data and of an algorithm in a computer (computer simulations).

Theoretical model, real experiment, and simulation are closely connected. Each simulation depends upon the knowledge of the biological reality. The biological facts are gained by experiments in conjunction with the formation of theoretical models and, perhaps, simulation studies. Then, the results of a simulation can be verified by real experiments and, finally, lead to further development and/or testing of theoretical models.

Conditions inserted into simulation studies are called parameters. The term system (simulator) includes a set of parameters arranged by an algorithm which reproduces a real population and its development.

II. PERFORMANCE OF SIMULATIONS

All simulation methods use the *Monte-Carlo* method. In this method, random numbers are generated from certain distributions. These numbers determine the occurrence of events which apparently take place in biological systems by chance (*e.g.*, chromosome distributions in meiosis, fusion of gametes, and environmental factors).

Simulation studies of selection acting upon quantitative traits are executed in digital computers. Full details of the technical part of computer simulations are given by Fraser and Burnell (1970), Bellmann and Ahrens (1973), and Crosby (1973).

In simulations of artificial selection, a population (initial population) is first generated with regard to theoretical models and biological facts. Starting with this population as generation 0, each of the following generations is iteratively generated using information from the previous generation. Thus, deterministic as well as random processes are simulated. By selection, individuals with certain character values or value combinations contribute disproportionately to the next generation. This causes a change of allele frequencies at the gene loci affecting the phenotype in question. The frequencies of linked alleles can be influenced by selection as well. The change of the allele frequencies affects the values of the characters and, in turn, affects the conditions of selection in the offspring.

After forming the initial population (generation 0), simulations of selection processes usually proceed along the following selection algorithm ($i = 1, \ldots, k$):

(1) Selection of parents for the next (ith) generation
(2) Combination of individuals for mating (mating system)
(3) Determination of the number of offspring (*e.g.*, per parent or parents)
(4) Production of gametes
(5) Combination of gametes (formation of zygotes, ith generation)
(6) Computation of genotypic and phenotypic values and so on, from generation to generation.

In a large number of simulations, the selection algorithm is repeated several times using the same initial parameter values. By such replicates random influences can be judged. Many repetitions should be chosen, if the influences of small populations, inbreeding, or environmental factors are to be considered. Moreover, many studies include several simulation runs with different parameter values or genetical models. Then the results are compared to gain information about the conditions and consequences of selection.

III. INCORPORATION OF BIOLOGICAL FACTS

In the published simulation studies, the biological facts are taken into consideration to varying extents. Simple systems save computer time and facilitate the interpretation of results. On the other hand, the results of simulations that model known genetic facts as far as possible come nearer to reality. In these more complex simulations, however, distinct results may no longer be explainable. This need not be a disadvantage, as such obscurities may lead to new questions.

In the simulation studies on artificial selection different biological conditions are incorporated. The number of gene loci which contribute to variation in a quantitative trait is always determined. In some papers only 1 or 2 loci are assumed; but, often between 3 and 100 loci are considered which is a range in accordance with some experimental evidence.

Simulations mainly look at diploid organisms. Only a few authors examine tetraploid conditions (Swanson et al., 1974a, 1974b). Sex-linked genes are seldom considered.

Linkages between gene loci and their map positions in the chromosomes are more or less realistically simulated. Some studies consider more unlinked gene loci than chromosomes in the eucaryotic haploid genome (e.g., Sather et al., 1977). Some published papers consider fixed rates of recombination between the genes, while other studies vary the degree of linkage between the gene loci.

In most studies only two alleles are assumed for each gene locus, though Bellmann and Ahrens (1973) give a method to simulate multi-allele cases. Most of the authors assign equal frequencies to all alleles in the initial population. Only in some simulations different initial allele frequencies at different gene loci are considered (e.g., Bellmann and Ahrens, 1973) or initial allele frequencies are varied between simulation runs (e.g., Hill and Robertson, 1966). In general, the allele frequencies at different gene loci are determined so that they satisfy the conditions of Hardy-Weinberg and linkage disequilibrium. Effects of non-chromosomal inheritance, which leads for example to maternal influences, are not included in any simulation study.

Intragenic effects are inserted as additive effects or as different degrees of dominance. In general, epistasis, when it is not neglected, is simplified as dualepistasis. Some studies (*e.g.* Gill, 1965a, 1965b, 1965c) use different kinds of dualepistasis. Bellmann and Ahrens (1973) try to meet the biological facts in a better way in that they simulate a complex system of intergenic relations.

In many papers environmental factors are considered as random variables which perturb the genotypic value of the zygote. No simulation study contains genotype environment interactions.

Most simulations consider only one trait. Bohren *et al.* (1966) were the first to investigate two characters with correlation caused by pleiotropy, while Singh (1967) and Singh *et al.* (1967) were concerned with correlation caused by linkage.

An equally-probable transmission of every homologous chromosome in the meiotic reduction cycle is generally assumed, so that all possible gametes contribute with equal probabilities to the formation of offspring zygotes. Concerning the number of offspring zygotes per parent, conditions of certain species are simulated. The number of parental individuals differs from population to population and is often varied in simulations.

Mass selection is used as the primary mode of selection. Sometimes selection is simulated within families (Sather *et al.*, 1977) or within lines (Bereskin, 1972). Special breeding systems, such as reciprocal recurrent selection (Cress, 1967; Ehdaie and Cress, 1973) or cyclic single cross selection (Ehdaie and Cress, 1973), have been investigated.

The intensity of selection is often varied. Selection pressure usually acts in only one direction. In some studies selection occurs simultaneously for several traits on the basis of indices (Sather *et al.*, 1977; Singh and Bellmann, 1974a, 1974b).

The number of generations in selection studies differs a great deal. It depends on technical factors (*e.g.*, computer time) or on the development of statistical quantities (*e.g.* reaching a plateau) which are calculated from the character values. Overlapping generations, as they exist in real populations, have not generally been simulated.

Mutations are not included in the simulations of artificial selection. But, under real conditions, mutations may be important for the selection advance, depending on the number of genes, the population size, and the number of generations.

IV. RESULTS OF SIMULATION STUDIES

The simulations of selection that have been carried out so far have produced some important results. The following parameters, which are shown in Table I with their initial values, are investigated in detail.

A. *Population Size*

For small populations and different models of gene action, Gill (1965a) described the influence of population size and selection intensity on the random fixation of alleles. For some gene action models, the study determined the smallest number of breeding individuals which was necessary to insure that no desirable alleles were lost by random drift. The influence of population size on the fixation rate of alleles was also simulated by Qureshi and Kempthorne (1968).

According to Bliss and Gates (1968) the genetic gain in small populations is lower than in large ones. This relation was also given by Qureshi *et al.* (1968). If the population size fell below a certain level, then, at a given recombination probability between the loci, the selection response was markedly slower. The authors stated that small populations together with close linkage strongly inhibit selection advance. In the case of overdominance they found a relation between allele fixation, selection pressure, and population size. In addition, a recent study (Sather *et al.*, 1977) investigates the influence of population size on the coefficient of inbreeding for different selection and mating systems.

B. *Selection Intensity*

The relations between selection intensity, selection advance, and the loss of desirable alleles were first examined by Gill (1965c). Also, Gill and Clemmer (1966) simulated the way an increasing selection intensity affected the coefficient of inbreeding in populations of distinct sizes. As shown by Young (1966), selection intensity mainly influences the decreasing additive genetic variance. The strong effect of selection intensity was also demonstrated in the simulations of Quershi (1968), Qureshi and Kempthorne (1968), and Qureshi *et al.* (1968). When selection became more intense, Parker *et al.* (1969, 1970a, 1970b) found decreasing values for the genetic correlation of two traits, one of which was selected for; this was primarily valid at large heritabilities and in the

TABLE I. Summary of the initial parameters in simulations of selection.[a]

Study[b]	Population size (unselected)	Selection intensity	Mode of selection	Allele frequency	Linkage	Gene Effects Intra	Gene Effects Inter	Number of traits	Environmental effects	Footnote
1	v	v	f	f (2/3)	v	f(d)	n.i.	1	n.i.	
2	v	f	f	f (0.5)	n.i.	v	v	1	n.i.	
3	v	v	f	f (0.5)	v	v	n.i.	1	v	
4	v	v	f	f (0.5)	v	v	v	1	v	
5		v	f	v	n.i.	f(a)	n.i.	2	f	c
6	v	v	f	f (0.5)	v	f(a)	n.i.	1	n.i.	
7	v	v	f	v	v	f(a)	n.i.	1	n.i.	d
8	f(1000)	v	f	f (0.5)	v	v	v	1	v	
9	+	+	v	v	n.i.	v	n.i.	1	f	e
10	f	f(10%)	f	f (0.5)	f	f(d)	n.i.	2	v	f
11	f(1000)	f(10%)	f	f (0.5)	v	f(d)	n.i.	2	v	f
12	v	v	v	f (0.5)	v	f(a)	n.i.	1	f	g
13	v	v	f	f (0.5)	v	v	n.i.	1	v	
14	v	v	f	v(0.1; 0.5)	v	v	n.i.	1	n.i.	
15	f	f	f	f (0.5)	n.i.	v	v	2	f	c
16	f(48) selected	v	f	f (0.5)	n.i.	v	n.i.	2	v	c

TABLE I. continued.

Study[b]	Population size (unselected)	Selection intensity	Mode of selection	Allele frequency	Linkage	Gene Effects Intra	Gene Effects Inter	Number of traits	Environ-mental effects	Foot note
17	f(500)	f(10%)	f	f v within model	v	f v within model	v	1	f	
18	+	+	v	v	n.i.	v	v	1	f	h
19			f	f (0.5)	n.i.	f(a)	n.i.	2	v	c
20			v	f (0.5)	v	f(a)	n.i.	1	f	
21	f(1000)	f(10%)	f	f (0.1; 0.5)	v	v	v	1	f	i
22	v	v	f	f (0.1; 0.5)	f	v	v	1	f	i
23	f(40)	v	f	f (0.5)	n.i.	f(a)	n.i.	1	v	j
24	v	v	v	f (0.5)	n.i.	f(a)	n.i.	2	f	k

[a] table abbreviations: f, parameter value fixed for all runs; v, parameter value variable; n.i., not included in the model; d, dominant; a, additive; +, special selection system.

[b] citations for studies: 1, Fraser, 1957; 2, Fraser, 1960b; 3, Martin and Cockerham, 1960; 4, Gill, 1965a, 1965b, 1965c; 5, Bohren et al., 1966; 6, Gill and Clemmer, 1966; 7, Hill and Robertson, 1966; 8, Young, 1966, 1967; 9, Cress, 1967; 10, Singh, 1967; 11, Singh et al., 1967; Singh and Bellmann, 1974; 12, Bliss and Gates, 1968; 13, Qureshi et al., 1968; Qureshi, 1968; 14, Qureshi and Kempthorne, 1968; 15, Bereskin et al., 1969; Bereskin, 1972; 16, Parker et al., 1969, 1970a, 1970b; 17, Bellmann and Ahrens, 1973; 18, Ehdaie and Cress, 1973; cont.

TABLE I. footnotes continued

[b] citations for studies, continued: 19, McMillan et al., 1973; 20, Bulmer, 1974; 21, Swanson et al., 1974a; 22, Swanson et al., 1974b; 23, Marani, 1975; 24, Sather et al., 1977.
[c] correlation by pleiotropy.
[d] entirely gametic system.
[e] reciprocal recurrent selection.
[f] correlation by linkage.
[g] self-pollinated plants.
[h] cyclic single cross selection.
[i] tetraploid.
[j] self-pollinated; number of loci variable.
[k] correlation by pleiotropy; different mating systems.

later generations of selection.

C. *Gene Linkage*

Fraser (1957) has demonstrated a marked influence of gene linkage on selection advance, though because of the strongly simplified simulation conditions, the value of his results is very limited. Under different conditions Gill (1965a) obtained the result that linkage had no influence on the change of allele frequencies and thereby on the genetic gain. Linkage disequilibrium, however, could bias the estimates of the components of genetic variance. Furthermore, Gill and Clemmer (1966) found that the development of the coefficient of inbreeding is influenced by linkage during selection. Hill and Robertson (1966) examined the relation between the degree of linkage and the probability of allele fixation; however, they made assumptions which strongly depart from real conditions. They described the effect of linkage by the product of $N \times c$ (effective population size times cross over rate). For large populations Young (1966, 1967) supported the results of Gill (1965a), confirming that linkage influenced neither the selection advance nor the proportion of genetic variance to total variance.

In their simulation study Singh *et al.* (1967) observed two characters, which were correlated by linkage. One character was selected, and recombination between it and the other locus was varied. They found that the correlation between characters became smaller with decreasing gene linkage.

Qureshi (1968) and Qureshi and Kempthorne (1968) investigated the influence of linkage on gene fixation probabilities and on the decrease of genetic variance. They got indications that close linkage delayed the fixation of genes. In this way the effect of close linkage on the inhibition of selection advance was partly neutralized. As Singh and Bellman (1974a) described, the effectiveness of selection indices is reduced by linkage. According to Bulmer (1974), the change of genetic variance during selection is determined by the harmonic mean of recombination fractions.

D. *Gene Effects*

Besides models with additive gene effects, several authors have simulated dominance models. Gill (1965a, 1965c) examined the relation between dominance, population size, and gene fixation. A study of Young (1966) stated that the prediction of genetic gain is more inaccurate with dominance than with additive gene action. According to Parker et al. (1970b), under complete dominance the direction of selection affected the genetic correlation between two pleiotropic characters differently, if there is selection for one of the two characters.

The influence of epistasis on character values in selected populations was discussed by Fraser (1960a, 1960b). Gill (1965a) examined gene fixation and inbreeding depression in selection processes for some epistatic models. In two further papers by the same author (Gill, 1965b, 1965c), the influence of epistasis was considered. According to Young (1967), the estimation of genetic gain becomes more inexact under the influence of epistasis. Moreover, the additive genetic variance seemed to increase under epistatic influence after some generations of selection.

E. *Correlations*

Simulation studies which considered more than one character were seldom realized. The first investigations come from Bohren et al. (1966) who examined the conditions for asymmetry of the correlated selection response of two pleiotropic characters. Singh (1967) and Singh et al. (1967) included two characters, the correlation of which was exclusively caused by gene linkage. In these papers a clear relationship is shown between gene linkage and the correlation of characters. Parker et al. (1970a, 1970b, 1970c) simulated the genetic correlation between two pleiotropic characters, one of which was selected. In their studies, selection reduced the correlation, while the size of heritability was more important than the intensity of selection. When dominance was present, the direction of selection was important for the development of the correlation (Parker et al., 1970b). Statements about the correlated response are made in a third paper (Parker et al., 1970c). Bereskin et al. (1969) and Bereskin (1972) selected for two pleiotropic characters using an index; thus, the relation between selection and inbreeding was a main aspect. Bereskin (1972) verified that limited inbreeding in selection programs which used intra-line-selection and cross breeding was concordant with considerable genetic gain. McMillan et al. (1973) considered pleiotropic characters. According to their paper, initially negative correlation between two characters decreases in the course of index-selection, if the selection pressure was nearly the same for both characters. Positive initial correlation moved towards 0. The effectiveness of different selection indices for correlated characters was compared by Singh and Bellmann (1974a, 1974b). They found that the effectiveness decreased if genes are closely linked, and they examined the influences of different parameter combinations (heritability, economical weights).

F. *Environmental Effects*

The simulation study of Martin and Cockerham (1960) demonstrated that the efficiency of selection declines with an increasing proportion of environmental variance. Different proportions of environmental variance to phenotypic variance were important for the genetic gain (Gill, 1965c), for the development of the additive genetic variance (Young, 1966), as well as for the correlation between pleiotropic characters (Parker et al., 1969). As Singh (1967) and Singh et al. (1967) reported, where the correlation is caused by linked genes, the magnitude of the correlated response of two characters does not depend on the heritability of the character under selection. Singh and Bellmann (1974a; 1974b) considered different heritabilities for characters under selection using an index. They found that the genetic gain per generation decreased more strongly when heritability was low.

V. SELECTION AND TARGET VARIABLES

In simulated systems, parameter effects are measured by different variables (target variables). Studies determine allele frequencies or deviations from allele frequency equilibria (*e.g.* inbreeding coefficient, fixation rates, linkage disequilibrium). The effect of selection is judged on the basis of means and genetic components of variance. If several characters are involved, correlations are given as well.

The behavior of target variables depends upon the simulated conditions. Systems and parameters are greatly simplified and generally differ among studies. This is the reason it is not possible to compare the results of different authors without describing the used conditions in detail. Therefore, one paper (Bellmann and Ahrens, 1973) was chosen for discussion because it simulates selection processes with only slight simplifications.

The authors simulated a quantitative character of maize (*Zea mays*). There already existed many experimental investigations, so that the genetical characteristics of the simulated system were sufficiently known. The choice of parameters in the initial population was as realistic as possible, including 10 chromosome pairs, 99 loci (3 genes of large effect

and 96 genes of smaller effect), different intragenic effects
(48 loci without dominance, 31 loci with partial dominance,
15 loci with complete dominance, and 5 loci with over-dominance), 30% of the gene loci with randomly assigned allele
frequencies in the initial population, heritability of 0.3,
and a population size of 500. Fifteen generations of selection were simulated with three variants of linkage and three
variants of epistasis. During selection the target variables
developed as summarized below.

A. *Allele Frequencies*

Generally, during the initial phase of selection, the
allele frequencies of genes of large effect converged more
rapidly towards fixation than did those of genes of smaller
effect. There was a direct correlation between the change of
allele frequencies of modifying minor genes and the magnitudes
of their modifying effects. In the early stages of the selection response, all genes which affect the quantitative character converged more rapidly than in the later stages.

During the initial phase of selection for alleles of large
effect, complete dominance results in a convergence which is
twice as fast as the convergence under partial dominance. In
later generations the situation was reversed. Under partial
dominance the allele frequencies for genes of large effect
changed independently of linkage. On the other hand, under
complete dominance the frequencies increased substantially
more slowly in the presence of linkage. This influence of
linkage disappeared if epistatically acting modifiers existed.
The fact that allele frequencies converged more quickly with
dominance than without dominance also held for genes of small
effect. If epistasis also occurred, however, the convergence
of allele frequencies was not stimulated by dominance.

In case of linkage, epistasis always hindered the convergence of allele frequencies for additive minor genes. During
the initial generations of selection the influence of linkage
was smaller than during later generations. Linkage could
cause the loss of desirable alleles for minor genes.

B. *Means*

For equal selection intensities, the values of the mean
increased more markedly under epistatic effects. The contribution of dominance to the raising of the mean during selection was negligible. Only very close linkage slowed down the
increase in the mean, but this was only true in the later

generations of selection.

C. *Variances*

As the allele frequencies of genes of large effect converged more quickly than those of genes of smaller effect, the genetic variance realized by major genes was reduced faster than the variance components caused by minor genes. Generally the development of genetic variance was primarily determined by the additive genetic variance; dominance variance was less important. During selection, epistatic variance decreased if the allele frequencies of the modifying genes converged towards 1 or 0, respectively, whereby genotypes with large combination effects dominated. In general, smaller additive genetic variances were observed in systems with epistasis, than in systems without epistasis, since genes of large effect were fixed more rapidly in systems with epistasis. The genetic variance was only slightly influenced by linkage. The whole genetic variance as well as the additive genetic variance decreased, independently of epistasis, to the same extent with and without linkage.

D. *Inbreeding Coefficient*

During selection the development of the inbreeding coefficient depended upon both linkage and epistasis. The inbreeding coefficient went up in populations with growing linkage. It increased faster in systems with more intense, than with slight, epistasis.

E. *Genetic Gain and its Prediction*

During a selection process in populations with slight or no linkage, the genetic gain decreased in the same way as the genetic variance. With close linkage the genetic gain diminished during the initial generations, though the genetic variance remained high. But after the initial phase of selection, the genetic gain rose slowly in populations with epistasis. This development corresponded to the epistatic variance.

The prediction of genetic gain was based upon the realized genetic variance (realized heritability). The realized heritability became less during the selection process, though it remained higher in populations with epistasis than in those without epistasis. With strong linkage the realized heritability decreased markedly. After 15 generations of selection the realized portion of additive genetic variance still amounted to 60% to 80% with or without slight linkage, respectively. With tight linkage the realized portion of additive genetic variance decreased to nearly 40%.

Comstock and Robinson (1952) proposed to use sib analyses for the prediction of genetic gain, and Bellmann and Ahrens (1973) examined the robustness of this statistical method. They found good agreement between predicted and observed genetic gain. Thus, estimates from sib analyses could explain the decrease of genetic gain, caused by linkage.

In experiments on maize (*Zea mays*) and rye (*Secale cereale*), the simulation results concerning the prediction of genetic gain could be verified (Bellmann and Griess, 1969).

VI. CONTRIBUTIONS OF SIMULATION STUDIES IN QUANTITATIVE GENETICS

For scientific aims a simulation study should be performed if the problem cannot be solved analytically or if the realization of an experiment is not possible or too expensive. Consequently, simulations can be valuable with complex questions as are found in connection with selection for quantitative characters.

The main limitation is that the transferability of the results of simulations to real populations depends on the extent to which the relevant biological processes are incorporated into the system. Therefore, a simulation study for a particular species requires minimum knowledge about the structures and effects of factors which influence the quantitative character under consideration. The choice of parameters, their values and simplifications are important. It should be possible to judge the simplifications rising from factors omitted or idealized. Moreover, investigators often neglect to choose realistic values for parameters in the initial population. On the whole, most of the cited simulations used severely idealized model populations. The important aim of simulating realistic situations was followed in only a few studies.

Perhaps a small restriction of the validity of simulation studies is based on the mathematical methodology. In many steps of the selection process random numbers are generated. Often, however, these so-called random numbers are in fact not really random, but correlated. If there are such correlations the results of a simulation can be faulty.

The interpretation of results is subject to limitations analogous to those of real experiments. In some simulation studies facts obviously included in the system are regarded as results afterwards. Besides, often very universal conclusions are drawn, although only some parameters with distinct values were considered in the simulations.

In principle, simulation studies offer some advantages in relation to real experiments:

(1) less expensive in material and time,
(2) control of more factors and/or exclusion of undesirable influences,
(3) almost total exclusion of systematic errors,
(4) repeatable under equal conditions,
(5) easier interpretation of results.

Simulation studies can be useful for one or more of the following aims:

(1) realization of experiments as a way of instruction,
(2) building and testing of models,
(3) examination of the robustness of a model or a statistical method,
(4) gain of theoretical statements from existing models,
(5) help to plan an experiment,
(6) stimulation to new experimental or theoretical investigations.

One contribution of simulations to the solution of scientific problems is the interpretation of data which were obtained by real experiments. In such cases several models are simulated to find a model which fits the data. The data from real experiments can also be used to prove the validity of a distinct model. To this aim the model is simulated and the results of simulation are compared with those of the real experiment. Moreover, the robustness of a mathematical-statistical method which is based upon a certain model can be examined by simulation studies. In other words, one can investigate the extent to which the validity of the mathematical-statistical method is limited when there are deviations from the assumptions of the model.

The simulation of models can make an important contribution to the deduction of new theoretical statements. The results can also serve to plan real experiments or to stimulate new considerations. As yet, these important tasks of simulations have seldom been realized in published work. It is, therefore, clear that the application and value of simulation studies will continue to grow.

REFERENCES

Bellmann, K., and Ahrens, H. (1973). *Nova Acta Leopoldina* No. 211, 38, 459-556.
Bellmann, K., and Griess, I. (1969). *Biometrische Zeitschrift* 11, 265-279.
Bereskin, B. (1972). *J. Animal Sci.* 34, 726-736.
Bereskin, B., Shelby, C.E., and Hazel, L.N. (1969). *J. Animal Sci.* 29, 678-686.
Bliss, F.A., and Gates, C.E. (1968). *Austalian J. Biol. Sci.* 21, 705-719.
Bohren, B.B., Hill, W.G., and Robertson, A. (1966). *Genet. Res.* 7, 44-57.
Bulmer, M.G. (1974). *Genet. Res.* 23, 281-289.
Comstock, R.E., and Robinson, H.F. (1952). In "Heterosis" (J.W. Gowen, ed.), pp. 494-516. Ames, Iowa State College Press.
Cress, C.E. (1967). *Crop Science* 7, 561-567.
Crosby, J.L. (1973). "Computer Simulation in Genetics." John Wiley and Sons, London.
Ehdaie, B., and Cress, C.E. (1973). *Theoret. Appl. Genet.* 43, 374-380.
Fraser, A.S. (1957). *Australian J. Biol. Sci.* 10, 492-499.
Fraser, A.S. (1960a). In "Biometrical Genetics" (O. Kempthorne, ed.), pp. 70-83. Pergammon Press, London.
Fraser, A.S. (1960b). *Australian J. Biol. Sci.* 13, 150-162.
Fraser, A., and Burnell, D. (1970). "Computer Models in Genetics." McGraw-Hill, New York.
Gill, J.L. (1965a). *Australian J. Biol. Sci.* 18, 599-617.
Gill, J.L. (1965b). *Australian J. Biol. Sci.* 18, 999-1007.
Gill, J.L. (1965c). *Australian J. Biol. Sci.* 18, 1171-1187.
Gill, J.L., and Clemmer, B.A. (1966). *Australian J. Biol. Sci.* 19, 307-317.
Hill, W.G., and Robertson, A. (1966). *Genet. Res.* 8, 269-294.
Marani, A. (1975). *Theoret. Appl. Genet.* 46, 211-231.
Martin, F.G., and Cockerham, C.C. (1960). In "Biometrical Genetics" (O. Kempthorne, ed.), pp. 35-45. Pergammon Press, London.
McMillan, I., Friars, G.W., and Quinton, M. (1973). *Genetics* 74, 169.
Parker, R.J., McGilliard, L.D., and Gill, J.L. (1969). *Theoret. Appl. Genet.* 39, 365-370.
Parker, R.J., McGilliard, L.D., and Gill, J.L. (1970a). *Theoret. Appl. Genet.* 40, 106-110.
Parker, R.J., McGilliard, K.D., and Gill, J.L. (1970b). *Theoret. Appl. Genet.* 40, 157-162.

Qureshi, A.W. (1968). *Theoret. Appl. Genet. 38*, 264-270.
Qureshi, A.W., and Kempthorne, O. (1968). *Theoret. Appl. Genet. 38*, 249-255.
Qureshi, A.W., Kempthorne, O., and Hazel, L.N. (1968). *Theoret. Appl. Genet. 38*, 256-263.
Sather, A.P., Swiger, L.A., and Harvey, W.R. (1977). *J. Animal Sci. 44*, 343-351.
Singh, R.K. (1967). *Int. Tgg. Math. Stat. Berlin 1966. Abhandl. Dtsch. Akad. Wiss. Nr. 4*, 143-149.
Singh, R.K., Bellmann, K. (1974a). *Theoret. Appl. Genet. 44*, 63-68.
Singh, R.K., Bellmann, K. (1974b). *Theoret. Appl. Genet. 44*, 289-293.
Singh, R.K., Bellmann, K., and Ahrens, H. (1967). *Biometrische Zeitschrift 9*, 240-249.
Swanson, M.R., Dudley, J.W., and Carmer, S.G. (1974a). *Crop Science 14*, 625-630.
Swanson, M.R., Dudley, J.W., and Carmer, S.G. (9174b). *Crop Science 14*, 630-636.
Young, S.S.Y. (1966). *Genetics 53*, 189-205.
Young, S.S.Y. (1967). *Genetics 56*, 73-87.

CANALISATION AND SELECTION

J. M. Rendel

CSIRO Division of Animal Production
North Ryde, NSW
Australia

I. THE CANALISED CHARACTER

Some characters are expressed to exactly the same degree in almost all members of a population. Uniformity of phenotype is most obvious when the character is countable, such as the number of fingers and toes in a vertebrate or the number of bristles on the scutellum of an insect. While a population can also be quite uniform with respect to characters that are not countable, such as the venation of an insect wing, uniformity for uncountable characters is not so easily measured.

Some countable characters seem to have a preferred mean without members of the population being noticeably uniform, *e.g.* the number of young born at a birth to a mammal, the clutch size in birds, and the number of small chaetae on the abdominal segments of *Drosophila melanogaster*. If a class in a frequency distribution is represented more often than expected, it is an indication that this class is canalised. Some characters which are obviously highly variable, such as the weight of a fly, may nevertheless depend on another for which the population is uniform. The weight of a *Drosophila* depends in part on the critical weight beyond which a larva can pupate successfully and up to which growth is exponential, and on the interval of time -- which is fixed for each of the species -- between the attainment of critical weight and pupation, during which growth is linear and depends on nutrition. It is the critical weight, the time to pupation, and the relationship between them which allows one species to grow larger than another on equivalent diets, and it is here that some uniformity within a population might be found (F.W.

Robertson, 1964; Royes and Robertson, 1964).

The uniformity of a population with respect to a canalised character is due to something more than the absence of genetic differences between its members. We are not concerned with the uniformity of homozygous inbred lines or genetically homogeneous clones and crosses between inbred lines, nor with equilibrium gene frequencies maintained by genetic homeostasis. We are concerned with the maintenance of uniformity of phenotype in the presence of variable genetic and environmental influences tending to deflect development from its normal path. The existence of genetic variation in a phenotypically uniform population has been demonstrated by treating a population in a way that exposes phenotypic variation of the canalised character and showing that the population responds to selection. Waddington (1952, 1953) treated a population of *D. melanogaster* to a heat shock of 40°C between the 21st and 23rd hours of pupal life; as a result 40% of the population failed to have a complete posterior crossvein. By breeding from flies with defective crossveins in one line and complete ones in another, he produced two populations, one of which failed to complete the posterior crossvein in 90% of the population, the other in 10%. Equally important was the appearance of flies with defective crossveins in the line selected for high incidence of failure even in the absence of treatment, demonstrating that the selected genotype operated on the developmental pathway in a quite general manner.

Mutant phenotypes are often highly variable by comparison with the wild-type; hidden genetic differences between members of a uniform population can be demonstrated by intro-

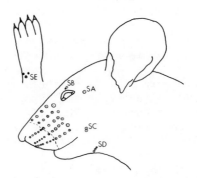

FIGURE 1. *Head and distal part of right fore limb of the mouse showing distribution of vibrissae. SE, ulnarcarpal; SB, supraorbital; SA, postorbital; SC, postoral; SD, interaural. (Redrawn from Fraser and Kindred, 1962).*

ducing a mutant into the population, selecting on the variations expressed in the mutant and demonstrating a genetic component by the response to selection. Such an experiment was carried out with the number of secondary mustacial whiskers in the house mouse (Dun and Fraser, 1959; Fraser and Kindred, 1960, 1962). This set of whiskers is arranged in four paired groups, left and right side, and one unpaired group below the chin (Figure 1). The total number is 19 with very few exceptions (Dun, 1959). The sex-linked gene *Tabby* reduces the number to a mean of 15 in $Ta/+$ and 8 in homozygous females and hemizygous males (Ta/Ta; Ta Y), with considerable variation about both means. Selection up and down is effective. Replacement of the Ta by the $+$ gene in the selection lines produces wild type mice with fewer than 19 whiskers in low lines and more than 19 in high lines. Selecting towards the mean restores the number of whiskers to 19 in $Ta/+$, but rarely do $Ta/+$ mice with more than 19 appear; the selected $Ta/+$ populations are canalised at 19.

An exactly parallel experiment was carried out using the mutant *scute* (*sc*; Figure 2) to reduce the number of scutellar bristles in *Drosophila melanogaster*. There are normally four bristles on the scutellum; the *sc* mutant has one or two, the number being highly variable. Selection towards four increases the number in *sc/sc* flies until the mean is well over three bristles with a large proportion of flies having four. At this stage a few $+/+$ sibs have five and even six bristles; but even though selection has increased the scutellar bristle number in the *sc/sc* population to the point where most flies have four bristles, few have more than four; *sc/sc* flies are canalised at four scutellar bristles (Rendel, 1959a). In flies homozygous for *ocelliless* in *D. subobscura* the number of ocelli and their accompanying bristles is reduced from the

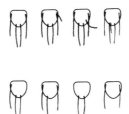

FIGURE 2. *Patterns of bristles on the scutellum of Drosophila melanogaster. The normal arrangement is shown in the top left, followed by typical examples of extra bristles and absent bristles. (Redrawn from Fraser, 1963).*

normal invariable three ocelli and eight bristles to a variable number which selection can restore to the normal eleven units; persistent selection eventually results in a population in which some flies have more than eleven units, but *ocelliless* flies, like wild-type, are canalised at three ocelli and eight bristles (Sondhi, 1961).

II. MEASUREMENT OF CANALISATION

The strength of canalisation can be measured by probit analysis (Finney, 1947). Canalised characters, which are characters with two thresholds, can be analysed as all or none characters (House, 1952; Rendel, 1959a; Latter, 1964). In this analysis it is assumed that a population has been classified on some scale, such as dead or alive or having 1, 2, 3, 4, 5 or 6 toes, and that the frequency of each class reflects an underlying variable that is normally distributed. The position of the boundaries of different classes can be related to the mean of the distribution of the underlying variable by referring the percentage of the population falling beyond the boundary to tables of the normal curve. If we know that 25% of a treated population died and 75% lived we can say that the boundary between death and survival lies 0.6745σ from the mean of the underlying variable that determines response to treatment, since in a normal curve a threshold cutting off 25% of the distribution lies 0.6745σ from the mean. The position of the upper and lower boundaries of all classes of a population relative to the mean of the underlying variable can be determined and, from these, the difference between the upper and lower boundaries of each class in standard deviations. This distance is referred to as the width of the class. The width of a canalised class is used as a measure of the strength of canalisation. Sample data are presented in Table I.

The width of a canalised class will be greater than the width of other classes in the population. Whereas the width of the uncanalised classes will depend mainly on the variance of the underlying variable, the width of the canalised class depends also on the efficiency of the mechanism that tends to keep development on a path leading to the canalised phenotype. The width of uncanalised classes has often been taken as a measure of the variance of the underlying variable (Rendel, 1959b; Sheldon, 1968; Latter, 1970). Objections to this use of the probit analysis (A. Robertson, 1965) are discussed below.

TABLE I. *Distribution of bristle phenotypes and probit analysis for females from a canalised and from an uncanalised population of scute flies. (From Rendel, et al., 1966).*[a]

	Bristle number				
	0	1	2	3	4
CANALISED POPULATION					
Number of flies	6	18	627	25	3
Cumulative percentage	0.85	3.40	92.21	99.57	100
Probit	2.61	3.18	6.42	7.58	
Width of class		0.57	3.24	1.16	
UNCANALISED POPULATION					
Number of flies	29	102	389	306	139
Cumulative percentage	3.01	13.57	53.88	85.60	100
Probit	3.10	3.90	5.10	6.06	
Width of class		0.78	1.20	0.96	

[a] *Isogenic lines taken from Oregon RC include one line, line 5, in which 6 males had three scutellar bristles and 8 had five out of a total of 8835 which gives a width of class 4 of 6.32±0.16; in line 20, 11 males had less than 4 bristles and 75 had more than 4 out of a total of 1412, which gives a width of class 4 of 4.06±0.12.*

(Quoted from Amer. Natur. 100 (910), 1966, by permission of the University of Chicago Press.)

III. THE CANALISING PHENOTYPE

In experiments cited so far, genetic sources of phenotypic change are the mutation of a major gene from its wild-type form, which reduces the mean score of the mutant population, and minor genes whose frequency is increased by directional selection. It was first supposed that canalisation was determined entirely by the major gene, *i.e.* by the genes at the *Tabby*, *scute*, and *ocelliless* loci (Dun and Fraser, 1959). The major gene was supposed to determine the shape of a sigmoid curve relating phenotype to genotype with a mean determined by the level of expression of the major gene. Mutation reduced the steepness of this curve and so increased phenotypic variance of the mutant no matter what its mean. But later results have made it plain that this is not so. It is at a precise phenotype that a character is canalised; it is at nineteen whiskers in the mouse, at four scutellar bristles in

D. melanogaster, and three ocelli plus eight bristles on the head of *D. subobscura*, and this is the case whether the wild-type or the mutant gene is present. It is necessary to consider a third genotype, sensitive to the course of development, which tends to direct it towards a precise end point. We may suppose that a genotype or environment or combination of the two that fails to reach the canalised phenotype is variable because it does not reach a level at which control operates. A combination of genotype and environment that goes too far is variable, because it has saturated the capacity of the control mechanism. Selection lines, up and down, in which the majority of the population still have the canalised phenotype have an upper or lower tail, as the case may be; in this tail the frequency of different phenotypes is what is expected in the tail of a normal distribution.

If canalisation has been the response of a species to natural selection acting on the advantages of having a particular phenotype and a regular pattern of development, we may expect to find some genotypic variation in the strength of canalisation in wild-type populations, not through mutation of the major gene but by variation of the canalising genotype itself. Isogenic lines extracted from a wild-type population of *D. melanogaster* will be samples of the genotype containing different sets of genes which may be expected to vary from line to line in the strength of canalisation at four scutellar bristles (Sheldon, *et al.*, 1964). The strength of canalisation in the parent stock from which isogenic lines were extracted was 5.4σ; the strength of canalisation in the eighteen isogenic lines extracted from it ranged from 6.4σ in the most strongly canalised to 3.4σ in the least. Genetic variation of scutellar bristle number within each isogenic line should have been largely eliminated by the isogenising process, and the width of the five bristle class did increase from 1.05σ to 1.15σ indicating a fall in variability. This is not sufficient, however, to account for an increase of 1.0σ in the strength of canalisation of the 4 bristle class in the strongest line, nor can any such explanation account for differences in canalisation between lines. A difference of 3σ between the strongest and weakest lines must imply a difference in the canalisation mechanism itself.

We can also ask whether it is possible for selection to introduce canalisation of a phenotype into the population at a level not previously canalised. The width of the two scutellar bristle class in *sc/sc* females is normally 1.3σ to 1.5σ. Fifty generations of selection increased the width of the 2 bristle class to 3.5σ. Further selection increased it to 4.0σ, but on relaxation it returned to 3.5σ. The width of the 2 class is always less in males (Rendel and Sheldon, 1960;

Rendel, *et al.*, 1966). In arithmetic terms 91% of females and 77% of males had two bristles instead of 52% of females and 28% of males. In a series of selection runs covering 30 generations, a population of *Ta/+* mice in a line segregating for *Ta* was canalised at fifteen whiskers; the strength of canalisation of the 15 class increased from 1.2σ to 2.2σ. At the same time, the 7 class in *Ta/Ta - Ta* Y mice increased from 1.3σ to 2.4σ. In a second line *Tabby* mice were selected at 7 and reached a level of 3.5σ from a level of 1.2σ; at the same time canalisation at 15 whiskers in *Ta/+* sibs in this line fell from 0.8σ to 0.4σ. In these selection lines there is the complication that total whisker number is a reflection of what happens at each of 5 whisker sites (see below) and that in *Ta/+* mice three sites behave like + and three like *Ta*. Selection in *Ta/Ta - Ta* Y has not only reduced the width of the 15 class in *Ta/+*, it has also reduced the mean and it will not be possible to interpret this result until one knows whether selection has changed the extent to which the *Ta* rather than the + X chromosome is active at whisker sites. The fact that some sites are predominantly like + and others like *Ta* could be accounted for on the inactive X hypothesis if there is some control over the choice of X to be inactivated. Selection in *Ta/+* for canalisation seems to have canalised the expression of the whisker characters at all sites in *Ta/Ta - Ta* mice except the ulnacarpal sites (Johnston *et al.*, 1970; Pennycuik and Rendel, 1977). Selection can also canalise wild-type characters at a new level. Scutellar bristle number, for example, has been canalised at six bristles (MacBean *et al.*, 1972).

Canalisation is the property of a specific phenotype; it is not readily achieved by selection routines designed to reduce variability in general. Selection was most effective when individuals with the desired phenotype were taken from families in which the desired phenotype was most frequent. This can be distinguished from centripetal or stabilising selection in which selection is against extremes without regard to the mean of any one generation. It can also be distinguished from selection which reduces the sensitivity of the mean to environmental and genetic influences without reducing variation about the mean. For example, the sensitivity to temperature of the size of the eye in the mutant *Bar* has been drastically reduced by selection in *D. melanogaster* (Waddington, 1960; Waddington and E. Robertson, 1966). The variation of the size of the *Bar* eye was not reduced by this selection though the difference between the mean at $18°C$ and $25°C$ was. Selection designed to canalise abdominal bristle number in *scute* flies at 7 bristles completely failed in 66 generations, but the difference between males and females fell from 3.9 (males, 5.6; females, 9.5) to 1.1 (males, 6.6; females, 7.7), the width of the 7 class remaining virtually constant at a little over 0.8σ (Rendel, 1969). In the mutant *Hairy wing* selection succeeded in decreasing the number of hairs in one cell of the wing in *Hw/Hw* of *D. melanogaster* to the same number as in *Hw/+* in one line and increasing the number in *Hw/+* to the same number as in *Hw/Hw* in another with relatively small effects on variance and without any preferred bristle number class appearing (Ohh and Sheldon, 1970). Druger (1967), who showed that selection can change the sensitivity of bristle number to temperature in *D. melanogaster* without changing variance, has discussed with further examples the genetic control over differences between means and differences about means.

IV. MECHANISM OF CANALISATION

It has been suggested that canalisation results from the control of a major gene in a developmental pathway. It was supposed that the major gene has to be turned on and can be turned off. It was further supposed that the canalisation genotype, which is sensitive to the progress of the developmental path in which the major gene acts, makes the molecules responsible for turning the major gene on and off, and does so to compensate for deficiencies and excesses in the elaboration of the major gene product (Rendel *et al.*, 1965; Rendel, 1969). The actions of minor modifying genes were supposed not to be controlled, at least not by reference to the progress of the developmental pathway in question. This suggestion is compatible with what is known about gene regulation. Some support for this idea comes from a comparison of the strength of canalisation in mutant and wild-type stocks. In unselected wild-type *D. melanogaster* stocks, the width of the four scutellar bristle class is 5.4σ in females. In a *scute* stock in which mean bristle number in females had been increased to 3.65, the width of the four class was 3.8σ (Rendel, 1976). In a population of *D. melanogaster* into which a weak allele of the + gene at the *scute* locus had been introduced, the width of the four scutellar bristle class fell from 5.4σ to 4.5σ (Scowcroft *et al.*, 1968). The width of the 19 whisker class in wild-type mice is 3.5σ. In *Ta/+* mice in selection lines in which the number of whiskers had been raised to 19, the width of the 19 class was 2.0σ (Fraser and Kindred, 1960; Kindred, 1962). There is an association between the strength of canalisation and the strength of the gene at the major gene locus. This suggests that control of the phenotype is exercised through the major gene, being stronger when the activity of the major gene that can be modulated is greater. The major gene changes the strength of canalisation but not the precise phenotype at which the character is canalised.

V. SELECTION IN *SCUTE* AND SELECTION IN WILD-TYPE

So far we have recognised two sets of genotypes in addition to the major gene in the *scute* developmental pathway; selection for one enhances the action of *scute* towards the normal phenotype and for the other canalises the *scute* developmental pathway. In *scute* populations attention is focussed on the single developmental pathway isolated by the *scute* mutant. In wild-type populations this is not so and it can be anticipated that sources of variation will be exposed in wild-type that would be overshadowed in *scute* populations. Further, whereas in *scute* populations there can be no selection pressure tending to decanalise the four bristle phenotype since it is not reached in any but highly selected *scute* populations, in wild-type populations any trend towards decanalisation will immediately come under pressure when selection is for increased bristle number.

Selection in *scute* populations has had limited effects. A population selected for many generations reached a mean in females of 3.85 bristles and was not increased beyond this point despite continued selection. When a wild-type gene was backcrossed into this line until the mean in *sc/sc* females was 3.65 bristles, the mean in +/+ was 4.20 (Rendel, 1976). This is an increase over unselected populations of about 2.0σ in *scute* and 1.2σ in wild-type. The reason for the difference is not established. It might be due to selection for stronger *sc* alleles or for genes close to *sc* on the X chromosome. Modifiers of *scute* are known to exist in this region (Whittle, 1969; Fraser, 1963; Payne, 1918) and would not have entered the X chromosome during backcrossing. Fraser (1966) found much the same in his *scute* selection lines. He reached means in *scute* females of 2.85 to 3.50; wild-type males backcrossed into the *scute* selection lines had means of from 4.07 to 4.30. It is clear that selection in *scute* populations does pick out genes that increase bristle numbers in wild-type, but the response to selection in *scute* seems to be limited. If a mean of 3.85 is the limit, modifiers of *sc* that increase the expression of the *sc* gene from 0 to 3.85 are exercising a total effect in high selection lines that cannot be more than 6.0σ. This can be compared with the substitution of *sc* by +, that is to say, the difference between $sc/+$ and sc/sc which is about 7.0σ (Rendel, 1959b) and the difference between $sc/+$ and $+/+$ which is 1.2σ. Selection in wild-type rapidly increases the scutellar bristle number by steps with plateaux of no response in between. Any or all of the three major chromosomes may be important. Interactions between selected chromosomes are varied and sometimes negative. Finally, selection in wild-type does drastically reduce canalisation at four scutellar bristles.

A number of workers have attained very high numbers of bristles by selecting in wild-type. Sismanidis (1942) reached 5.2 in females in 24 generations of selection. Most of the increase was due to the second chromosome in the early generations. Payne (1918) reached ten bristles in 35 generations with plateaux at 6, 8, and 9 bristles. He found that the first and third chromosomes were the ones responsible for increasing bristle number. Fraser and his co-workers (Fraser, 1963; Fraser et al., 1965) reached a mean of 8.5 bristles in 50 generations with plateaux at 5 and 7 bristles. In these lines it was the first and third chromosomes that had responded (Fraser and Scowcroft, 1965; Scowcroft, 1966). Whittle (1969) found that in one line that had reached 9.25, the first and third chromosomes were almost entirely responsible for the increase, the second having a relatively small effect, whereas in a second line reaching 8.05 most of the increase was due to the second chromosome. MacBean et al. (1971) who reached 16 bristles in 90 generations found at the 10 bristle stage that the second chromosome had the largest influence, the first and third had about the same influence as each other and considerably less than that of the second chromosome. Sheldon and Milton (1972) in one line reached a mean of 8 in 15 generations, 10 in 50, and 13 in 90 with pauses at each step. Other lines responded in a different pattern, but all reached high means. The contrast between increases in bristle number as a result of selection in wild-type, which reaches means of 13 and more, and selection in *scute* which reaches means in wild-type of 4.1 to 4.3 is clear.

VI. THE EFFECT OF SELECTING IN WILD-TYPE ON *SCUTE* POPULATIONS

The contrast between selecting in *scute* and wild-type populations suggests that at least some of the genotype on which selection operates differs in the two genotypes.

Selection in wild-type can increase bristle number in *scute* sibs as was shown by Sheldon and Milton (1972) who backcrossed *scute* into two of their selection lines. After five generations of backcrossing, wild-type flies were 5.6σ above the unselected mean in one of these lines, *sc* and *sc/sc* flies being up by 2.9σ and 2.3σ, respectively. In a second line wild-type flies were 3.6σ above the unselected mean and *scute* flies were up by 2.3σ. On the arithmetic scale this corresponds to means in females of 8.1 bristles instead of 4 in wild-type and 3.21 instead of 1.2 in *scute* in one line; and 6.08 instead of 4 in wild-type and 3.1 instead of 1.2 in *scute* in the other line. Although the wild-type means are not as high in the backcross as in the parental wild-type line into which the backcross was made, they are much greater than those of 4.07 to 4.3 resulting from selection in the presence of *scute*, even though the means of the *scute* sibs in the backcross line are of the same order or somewhat less than the means achieved in the *scute* selection lines. Clearly there are genotypes in wild-type which increase wild-type bristle number, but which are not expressed in *scute* lines.

Fraser and his coworkers have emphasized the distinction between selection in wild-type and *scute*. In order to examine this point, *scute* was backcrossed into a variety of their wild type selected lines. Y chromosomes bearing the tip of the X, including the *scute* locus, were also introduced. These lines segregate for a variety of genotypes in each line, both with and without *scute*, and with and without the extra *scute* locus attached to the Y. The effect of the background genotype in the selection line on this variety of genotypes was then compared, and it was found that there was a correlation between two genotypes if both had means above 4 and if both had means below 4, but no correlation between pairs of genotypes if one had a mean above and the other a mean below 4 (Fraser and Green, 1964). They concluded that two sets of modifiers were present, one set increasing bristle number in *scute* flies and one in wild-type, and that these two sets had been increased independently in their wild-type selection lines. A gene, *extravert*, located on the third chromosome at about 41cM and possibly allelic to *polychaetoid* was found in one line. This gene which is recessive markedly increased scutellar bristle number in the wild-type selection lines in which it was found as well as increasing the number of vertical bristles (Fraser, 1967; Miller and Fraser, 1968; Fraser, 1970). This gene, together with other wild-type modifiers, is not expressed in *scute* flies. However, in related selection lines which did not carry *extravert*, but presumably did contain many of the modifiers present in the *extravert* lines, both *scute* and wild-type had elevated bristle numbers. It was necessary, therefore, to suppose not only two sets of modifiers, the α set affecting *scute* and wild-type and the β set affecting wild-type only, but also to suppose that *scute* inhibited *extravert* and β modifiers and that *extravert* inhibited α modifiers.

VII. REDUCTION OF CANALISATION AS A RESULT OF SELECTION IN WILD-TYPE

In a carefully analysed series of experiments, Latter (1964, 1966, 1970, 1973) found that, on the probit scale, response to selection was steady with a constant heritability of 40% and a realised heritability of 34% until the mean reached 7.5. From then on, response was steady but at a heritability of 23% and a realised heritability of 10%. At the 8/9 threshold he found a ceiling, approach to which was accompanied by a steady decrease in the influence of genes of all chromosomes and a decrease in fitness. Beyond this threshold all chromosomes contributed, the third chromosome being the most important. In one line (SH.1) scutellar bristle number reached a mean of 13. He found that it was the third chromosome in this line which took it beyond the 8/9 ceiling. The most powerful region of this third chromosome was located at 61cM near *sr* and was found to add 5 bristles. Three other regions were also located which added from 3.0 to 0.8 bristles. The region near *sr* was recessive and increased bristle number in those areas normally affected by *achaete* as well as increasing the number of scutellars. Latter (1973) was able to demonstrate that the standard deviation of scutellar bristle number increased two-fold in lines crossing the 8/9 threshold. Up to this point there appeared to be no increase in variance in any of his lines. When a weak wild-type gene was introduced into two of his selection lines (SB9, which had plateaued at the 8/9 barrier, and SH1, which had reached a mean of nearly 13), bristle number was reduced so that means were 3.92 and 5.07, respectively, and the width of the 4 class could be measured directly. In SB9 the width was 3.38σ and in SH1 1.70σ, showing that the increase in variance found beyond the 8/9 barrier was accompanied by a decrease in canalisation. No direct comparison could be made with the width of the 4 class in an unselected population, but it was possible to compare the width of the one bristle per site class at anterior sites. This was 4.23σ in the unselected population, 4.50σ in SB9 and 2.94σ in SH1 (Scowcroft and Latter, 1971). Clearly SB9 and SH1 differ in the extent to which the 4 scutellar bristle class is a preferred class and the authors conclude that selection has reduced the power of regulation in the SH1 line. Rendel

(1976) isolated the gene at 61cM on the third chromosome which he called r^1 in a region marked by *cu* and *sr*. It was not possible to measure the width of the 4 class in wild-type populations carrying *cu*, r^1, *sr* as the number of scutellar bristles was too high. However, when backcrossed into a high selection line carrying *sc*, it reduced the width of the 4 class from 3.81σ to 1.94σ. So this small chromosome segment reduces canalisation in *scute* flies if the mean in the *scute* stock is sufficiently high. As it is the strongest element in Latter's third chromosome which decanalises wild-type, there is *prima facie* evidence that the same gene decanalises both wild-type and *scute* populations and that it is primarily the phenotype with which the genotype for decanalisation interacts.

Sheldon (1968) and Sheldon and Milton (1972) found that canalisation measured by the width of the 4 class decreased almost as soon as selection began to take effect. Three selection experiments starting with wild-type flies have demonstrated that increases in bristle number were accompanied by a decrease in canalisation. If the mechanism of canalisation suggested above is operating, the increase in bristle number in wild-type populations which is not found in *scute* populations would have resulted from the partial liberation of the major gene from control. A genotype which decanalises the character will not act in an unselected *scute* population with a bristle number so low that the canalising mechanism is never activated.

According to this hypothesis, an increase in suppression of the *scute* locus decreases bristle number to 4 but not beyond; a decrease in suppression increases bristle number above 4. Fraser and Green (1964) suggested two sites of control instead of one, an enhancer and an inhibitor. It is not clear how these sites are related to α and β modifiers.

VIII. WHAT IS IT THAT IS BEING CANALISED?

Alan Robertson (1965) has drawn attention to the dangers of drawing conclusions on the assumption that the classes into which the individuals in a population are placed for the purpose of probit analysis reflect the distribution of some underlying variable. He refers particularly to four bristles on the scutellum of *Drosophila melanogaster* and suggests that a frequency distribution showing how many individuals in the population have 3, 4, 5, 6, or more bristles on the scutellum may not reflect the distribution of a variable underlying

bristle number as a whole, but rather the expansion of the
probability that there will be 0, 1, 2, or more bristles at a
bristle site. He points out that the frequency distributions
generated in this way approximate those actually found. If he
is correct, canalisation is the result of increasing the probability of one and only one bristle at a site and the width
of classes other than the 4 class do not reflect the variance
of the underlying variable, but reflect the probability of
more or less than one bristle at a site. He supposes that
whatever the mechanism of canalisation is, it is not sensitive
to what happens over the whole scutellum but only to what
happens at a bristle site. If all sites are independent it
should be possible to predict from the frequency of 0, 1, 2,
or more bristles at a site what the frequency distribution of
flies with 0, 1, 2, 3, 4, 5, or more bristles is going to be.
A prediction on this basis was compared with actual frequency
distributions in four *scute* stocks (Rendel, 1965) and found to
be a poor fit. In particular, a number of flies with 5 bristles was predicted, but none found. In a stock canalised at
2 bristles the correlation between anterior and posterior
bristles was -0.6, far from independent. Finlay (1965) selected in this stock for flies with one anterior and one posterior bristle on the opposite side and in ten generations increased their frequency from 12% to 17% in females and from
14% to 26% in males without reducing the strength of canalisation. There seems to be some justification for supposing that
there is regulation of bristle number over the scutellum as
a whole. In contrast, Pennycuik and Rendel (1977) found that
total whisker number in mice could be predicted accurately
from whisker number at each of the five sites where secondary
mustacials grow and concluded that whisker number is regulated
separately at each site, a site in the mouse corresponding to
the scutellum in *Drosophila*. In *Drosophila* canalisation of
the scutellar bristle number at 2 has no canalising effect on
the other sites at which *sc* reduces the mean (Rendel and
Evans, in press).

A distinction has been made between selection in *scute*,
which restores the normal number of bristles but which does
not increase the number of sites at which bristles appear, and
selection in wild-type, which not only increases the number of
bristles at a site but also increases the number of sites. In
addition to extra anterior and posterior bristles, bristles
are found between the normal anterior and posterior sites, not
associated with either. This contrast must be considered in
relation to the distinction drawn by Stern (1954) between prepattern (*e.g.*, the establishment of a bristle site) and the
activation of the site to form a bristle. He made flies with
mosaic patches of *yellow achaete* tissue and showed that where

ever a patch of *achaete* tissue, however small, covered an "*achaete*" bristle site, no bristle was formed. The same has been shown to be true of *sc* (Sturtevant, 1932; Young and Lewontin, 1966). When a patch of *sc* tissue in an otherwise normal scutellum occupied a bristle site, no bristle was formed. The converse is also true; when a patch of wild-type tissue occupies a bristle site in an otherwise *sc* fly a bristle is formed (Young and Lewontin, 1966). Thus, the gene at the *scute* locus is autonomous. It is also strictly localised at each site. Only one cell normally differentiates into a bristle-forming cell together with the collar cell, the sense organ, and nerve cell which go with it, and there are only four such cells in the whole scutellum.

Following Stern's idea of a prepattern, we must suppose that at least two processes are at work. Cells on the scutellum become determined as potential bristle-forming cells; then a signal resulting in the differentiation of these cells is elaborated by the gene at the *scute* locus. Whether the *scute* gene is active across the scutellum and only determined cells respond or whether the *scute* gene is turned on only in determined cells is not clear. We can suppose that the cells that are determined are much more likely to differentiate than their neighbors. In the absence of any change in the number and pattern of determined cells, extra bristles appear near normal bristle sites when the signal to differentiate spreads with sufficient strength to differentiate cells next to the normal bristle cell. This sort of response is the one expected to follow from selection in the presence of *sc* and as a result of adding extra doses of the *scute* locus (see below). Extra bristles will also be formed if the number of determined cells is increased and their distribution extended. Since r^1, which increases the number of sites in wild-type, does not appear to do so in *scute* we must suppose that the cells at the new sites require stronger signals to differentiate than the cells at normal sites. There would be two sorts of canalisation according to this hypothesis, one regulating the strength of the signal to differentiate, the other the number and sensitivity of determined cells. Maynard Smith and Sondhi (1960) and Maynard Smith (1972) have discussed the formation of prepatterns by gradients on the lines suggested by Turing (1952).

IX. DOSAGE COMPENSATION AND SELECTION

The mutant *scute* is a hypomorph since wild-type levels of expression are approached as the number of doses of the *sc* gene are increased. The effect of adding extra doses of the *scute* locus is controlled in such a way that the greater the bristle number in the stock to which the extra locus is added the smaller the effect until a point is reached at which the equivalent to between 2.0 and 2.5 bristles is added. At this point the number of bristles added becomes constant (Rendel et al., 1965). This control has not as yet been overridden by selection to produce high wild-type lines. Even very high lines are only increased by 2.0 to 2.5 bristles by the addition of an extra *scute* locus (Sheldon and Milton, 1972). There is, thus, one potential source of bristle variation still untapped by selection procedures in the developmental pathway leading to the formation of scutellar bristles.

X. CONCLUSIONS

The number of bristles in *D. melanogaster* and the number of whiskers in mice are superficial characters and the end products of development; the outcome of their development is directly observable. It is likely that the development of other characters, not so readily observed because they are not superficial and having consequences in subsequent steps in development, will follow similar patterns. The features of variation in *scute*, *Tabby*, and *ocelliless* are not readily related to the simple statistical models of quantitative genetics; they will not be fully understood without a detailed analysis of development. They suggest that much of the variation observed in populations, particularly rare phenodeviants, is concerned with the regulation of very powerful genes which are normally kept under strict control. The contribution of classical modifiers is relatively small by comparison with the potential of these major genes.

REFERENCES

Druger, M. (1967). *Genetics 56*, 39-47.
Dun, R.B. (1959). *Australian J. Biol. Sci. 12*, 312-330.
Dun, R.B., and Fraser, A.S. (1959). *Australian J. Biol. Sci. 12*, 506-523.
Finlay, D.E. (1965). *Amer. Nat. 99*, 431-438.
Finney, D.J. (1947). "Probit Analysis." Cambridge Univ. Press, Cambridge.
Fraser, A. (1963). *Genetics 48*, 497-514.
Fraser, A. (1966). *Australian J. Biol. Sci. 19*, 147-155.
Fraser, A. (1967). *Genetics 57*, 919-934.

Fraser, A. (1970). *Genetics* 65, 305-309.
Fraser, A., and Green, M.M. (1964). *Genetics* 50, 351-362.
Fraser, A., and Kindred, B.M. (1960). *Australian J. Biol. Sci.* 13, 48-58.
Fraser, A., and Kindred, B.M. (1962). *Australian J. Biol. Sci.* 15, 188-206.
Fraser, A., and Scowcroft, W.R. (1965). *Australian J. Biol. Sci.* 18, 851-859.
Fraser, A., Scowcroft, W., Nassar, R., Angeles, H., and Bravo, G. (1965). *Australian J. Biol. Sci.* 18, 619-641.
House, V.L. (1952). *Genetics* 38, 309-327.
Johnston, P.G., Pennycuik, P.R., and Rendel, J.M. (1970). *Australian J. Biol. Sci.* 23, 1061-1066.
Kindred, B. (1962). *Genetics* 48, 621-632.
Kindred, B.M. (1967). *Genetics* 55, 636-644.
Latter, D.B.H. (1964). *Genet. Res.* 5, 198-210.
Latter, D.B.H. (1966). *Genet. Res.* 8, 205-218.
Latter, D.B.H. (1970). *Genet. Res.*, 15, 285-300.
Latter, D.B.H. (1973). *Genetics* 73, 497-512.
MacBean, I.T., McKenzie, J.A., and Parsons, P.A. (1971). *Theoret. Appl. Genet.* 41, 227-235.
MacBean, I.T., McKenzie, J.A., and Parsons, P.A. (1972). *Theoret. Appl. Genet.* 42, 12-15.
Maynard Smith, J. (1972). "On Evolution." Edinburgh Univ. Press, Edinburgh.
Maynard Smith, J., and Sondhi, K.C. (1960). *Genetics* 45, 1039-1050.
Miller, D.H., and Fraser, A.S. (1968). *Australian J. Biol. Sci.* 21, 61-74.
Ohh, B.K., and Sheldon, B.L. (1970). *Genetics* 66, 517-540.
Payne, F. (1918). *Proc. Nat. Acad. Sci.* 3, 55-58.
Pennycuik, P.R., and Rendel, J.M. (1977). *Australian J. Biol. Sci.* 30, 303-317.
Rendel, J.M. (1959a). *Evolution* 13, 425-439.
Rendel, J.M. (1959b). *Australian J. Biol. Sci.* 12, 524-533.
Rendel, J.M. (1965). *Amer. Nat.* 99, 25-32.
Rendel, J.M. (1969). *Proc. Nat. Acad. Sci.* 64, 578-583.
Rendel, J.M. (1976). *Genetics* 83, 573-600.
Rendel, J.M., and Evans, M.K. (1978). *Heredity*, in press.
Rendel, J.M., and Sheldon, B.L. (1960). *Australian J. Biol. Sci.* 13, 36-47.
Rendel, J.M., Sheldon, B.L., and Finlay, D.E. (1965). *Genetics* 52, 1137-1151.
Rendel, J.M., Sheldon, B.L., and Finlay, D.E. (1966). *Amer. Nat.* 100, 13-31.
Robertson, A. (1965). *Amer. Nat.* 99, 19-23.
Robertson, F.W. (1964). *Genet. Res.* 5, 107-126.

Royes, W.V., and F.W. Robertson. (1964). *J. Exp. Zool. 156*, 105-136.
Scowcroft, W.R. (1966). *Genetics 53*, 389-402.
Scowcroft, W.R., and Latter, D.B.H. (1971). *Genet. Res. 17*, 95-101.
Scowcroft, W.R., Green, M.M., and Latter, D.B.H. (1968). *Genetics 60*,
Sheldon, B.L. (1968). *Australian J. Biol. Sci. 21*, 721-740.
Sheldon, B.L., and Milton, M.K. (1972). *Genetics 71*, 567-595.
Sheldon, B.L., Rendel, J.M., and Finlay, D.E. (1964). *Genetics 49*, 471-484.
Sismanidis, A. (1942). *J. Genet. 44*, 204-215.
Sondhi, K.C. (1961). *J. Genet. 57*, 193-221.
Stern, C. (1954). *Amer. Scientist 42*, 213-247.
Sturtevant, A.H. (1932). *Proc. 6th Int. Congr. Genet. 1*, 304-307.
Turing, A.M. (1952). *Phil. Trans. Roy. Soc. London, ser. B, 237*, 37-72.
Waddington, C.H. (1952). *Nature 169*, 278-279.
Waddington, C.H. (1953). *Evolution 7*, 118-126.
Waddington, C.H. (1960). *Genet. Res. 1*, 140-150.
Waddington, C.H., and Robertson, E. (1966). *Genet. Res. 7*, 303-312.
Whittle, J.R.S. (1969). *Genetics 63*, 167-181.
Young, S.S.Y., and Lewontin, R.C. (1966). *Genet. Res. 7*, 295-301.

THE POSTERIOR CROSSVEIN IN *DROSOPHILA*
AS A MODEL PHENOTYPE

Roger Milkman

Department of Zoology
The University of Iowa
Iowa City, Iowa

I. INTRODUCTION

Phenotypic variation is characteristic of species and essential to Darwinian evolution. Nevertheless, the genetic basis of natural phenotypic variation is still understood only in the broadest terms. A timely goal, therefore, is the explicit genetic analysis of a variable property at the appropriate phenotypic level. The appropriate level is one that is most directly scrutinized by natural selection and most easily investigated; thus, external morphology observable with nothing more than a dissecting microscope is suitable.

In this realm, the trait chosen need not be one whose variation has demonstrable adaptive significance. It would suffice to have a tractable genetic basis that is evidently similar in general ways to the genetic basis of adaptively significant variation in other traits.

Finally, the trait of choice must be easy to study in ways that can permit the explicit characterization, including mapping, of the loci whose polymorphism causes it to vary.

In over a century of study of natural phenotypic variation, no single trait has gained general recognition as a paradigm. We understand that genes do not usually act independently in evolution, and therefore that simple one-locus traits (*e.g.*, albinism) are not sufficient models. A trait with a somewhat higher degree of complexity, in which several loci are involved, may be both sufficiently representative and sufficiently tractable to provide a paradigm. In this chapter I will describe one such trait, together with two recent relevant technical advances.

How should a phenotypic paradigm of this sort be analyzed? The genetic variation required to span the detectable phenotypic range (a range that can be expanded by the use of environmental variables to balance genetic effects) should be defined. In the example to be presented, this comprises 3 to 5 segregating loci. Then the total set of relevant segregating loci in the species, from which the 3-5-locus-subsets are drawn, should be enumerated and characterized. As will be seen, this may be less overwhelming than it sounds.

II. DESCRIPTION AND CAUSALITY OF CROSSVEIN DEFECTS

The posterior crossvein is a simple linear strut. It runs between the fourth and fifth longitudinal wing veins of all members of the genus *Drosophila*, as well as many related species (Figure 1). Homologous structures exist throughout the Diptera.

A. *Quantitative Description*

Defects (missing portions) of the crossvein are sporadic and small in natural populations of many *Drosophila* species. They are uniformly extreme in various simple mutant strains and selected polygenic strains of *D. melanogaster*. It is possible to quantify the defects by dividing each posterior crossvein into imaginary sixths (Figure 1). Since there is a posterior crossvein on each wing, a range of 0-12 is obtained, with 0 representing the normal phenotype (no defects) and 12 representing the complete absence of the crossveins. A rating of 6 is given a fly with two crossveins each rated "3" (3/6 of each posterior crossvein missing), or a "4" and a "2", a "5" and a "1", or even a "6" and a "0". A smaller, anterior crossvein is also present in *Drosophila*. It will be ignored in the present discussion, in which "crossvein" will refer specifically to the posterior crossvein.

B. *An Underlying Variable and Thresholds of Expression*

These ratings are related to an underlying continuous variable, called "crossvein making ability", by two thresholds (Figure 2). A normal fly has a considerable reserve: this must be exhausted before crossvein defects can be produced. For example, one can produce crossvein defects by exposing

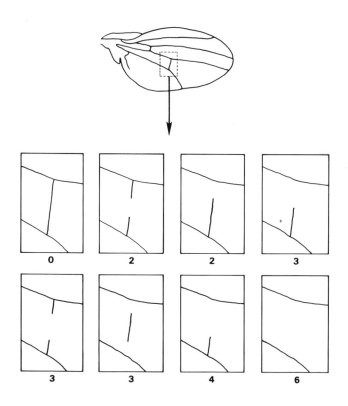

FIGURE 1. The posterior crossvein in Drosophila melanogaster and examples of defect ratings.

day-old pupae to a particular high temperature. There is a considerable sub-threshold range of durations which produce no effect. Longer durations produce increasing effects. Even longer durations kill the pupae, so that a second threshold is not seen in this case, but it *is* seen when the genotype is altered progressively. If, instead of increasing durations of exposure to high temperature, one substitutes increasing numbers of "*crossveinless*" (*cve*) alleles at any of several loci, again a certain number must be added before defects are seen. Then more *cve* alleles cause greater defects, until the

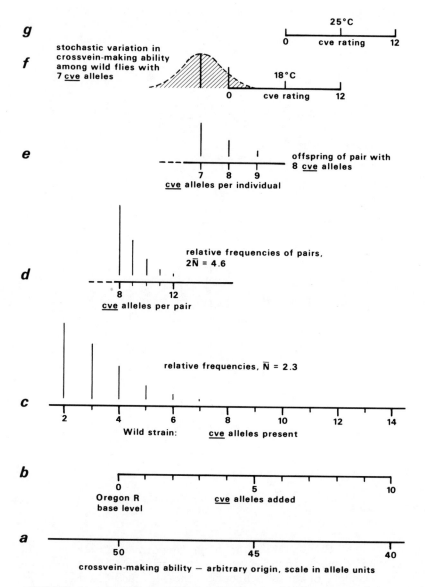

FIGURE 2. A scheme which unifies, starting at the bottom, the following properties in terms of corresponding values: (a). An underlying variable, crossvein-making ability. (b). Effects of cve alleles added to a standard inbred strain, Oregon R. (c). Genotypic variation in a wild population or in a strain derived from a wild population without a severe genetic bottleneck. (d). Distribution of cve alleles among pairs

posterior crossveins are totally absent. Even more *cve* alleles can be substituted, but no further crossvein defects are possible, so a second threshold is crossed.

How does one know that the last *cve* alleles to be added are capable of influencing the crossveins? One way is to add them in a different order, so that they come in between the thresholds. Another way is to shift the thresholds. For example, it takes more *cve* alleles to produce defects at $25°C$ than at $18°C$ (Figure 2, top).

Posterior crossvein defects constitute a quantitative trait with a threshold of expression. Such traits are common. Thresholds abound in biology, and development is no exception. A quantitative range of phenotypic expression is also common, and the combination of both quantitative variability and a threshold of expression is especially useful for the study of natural variation.

The quantitative aspect of a phenotype is important, because much natural variation is genetically collaborative: alleles at several loci contribute effects, and their number in a particular case can be estimated from the phenotypic value. The threshold is important, because it permits nearly effortless selectivity on the part of the investigator. Instead of measuring each individual and assigning one's own arbitrary threshold, the investigator simply observes which individuals have crossvein defects and which do not.

C. *Crossveinless Individuals as Phenodeviants*

The sporadic occurrence of deviant phenotypes, such as crossvein defects, due to a rare combination of individually common genetic and perhaps environmental variables had led to the term *phenodeviant*. Lerner (1954) proposed this term, not for a rare trait, but for an individual possessing a rare trait. No simple corresponding term exists for the trait itself, and "deviant phenotype" will have to do. Other examples are cleft lip (and palate), diabetes, and (in fowl) crooked toes. While Lerner originally suggested that phenodeviants lack the amount of heterozygosity necessary for normal development, he almost immediately modified this stand (personal communication) in view of the normal phenotype of highly homozygous strains of *Drosophila* and other organisms. A deviant

(FIGURE 2 continued) ancestral to isofemale lines, and (e). in the post-F_1 offspring of one such line. (f). Phenotypic rating at $18°C$, together with stochastic variation. (g). Phenotypic rating at $25°C$.

phenotype characteristic of a phenodeviant is now attributed to a rare combination of individually common alleles, recognizing that environmental factors may be of importance also. Nevertheless, phenodeviants are not associated with rare environmental extremes, nor with rare simple alleles. Albinos, for example, are not phenodeviants. The distinction is that albinism is a by-product of the ineluctable process of *mutation* to deleterious alleles; phenodeviants are the by-product of the process of *recombination* of alleles that are favorable in most combinations.

D. *Genetic Assimilation of the Crossveinless Phenotype*

Often the environmental variation that determines the threshold of expression of an underlying variable (like crossvein-making ability) has greater dispersion than the relevant natural genetic variation. Thus, certain environments may result in a threshold exceeded by all individuals, and other environments may place the threshold where it is exceeded by none, so that a trait may be associated primarily with a particular environmental factor. Under intermediate conditions where the range of the underlying variable straddles the threshold, it may be possible to select those alleles favoring the phenotype initially associated with a certain environment, so that eventually a genotype is assembled that formerly was vanishingly rare. Now the environment is no longer decisive; it may be of no consequence, and we call the process of change *genetic assimilation*. The posterior crossvein is once more the paradigm: Waddington's classic paper (1953) described the production of crossvein defects in a normal *D. melanogaster* stock of recent natural origin solely by pupal heat treatments. Those susceptible were selected, and eventually a true-breeding, polygenic strain was obtained with uniformly extreme defects regardless of the (non-lethal) environment.

III. A QUANTITATIVE MODEL

The following simplified numerical example is representative of the genetic and environmental influences on the posterior crossvein (Figure 2), with the following qualifications: We do not know how many relevant loci are actually polymorphic in a given population; and it is not true that the phenotypic consequences of all *cve/non-cve* allelic differences are exactly equal. Finally, we cannot make a firm general statement that all *cve* effects on crossvein making ability are precisely additive. We do know that the fixation of *cve* alleles at between 3 and 5 loci makes the difference between some *cve* strains and their normal ancestral populations; numerous and detailed studies on particular *cve* strains also support the general choice of values, though considerable variation doubtless exists. The loci that are segregating vary from strain to strain: the total number from which the effective sets are drawn is a question of major interest.

A. *Genotypes of Inbred Wild Strains and Panmictic Wild Populations*

In Figure 2, the standard *D. melanogaster* wild-type inbred strain, Oregon R, is depicted as having a base level of *cve* alleles designated "0". Since Oregon R is considered to be genetically uniform, there are no *cve* alleles segregating. At some *cve* loci, *cve* alleles may be fixed (monomorphic), and at others *non-cve* alleles are fixed. The position of the Oregon R base-level genotype is determined by the number of *cve* alleles that must be added to produce a given degree of crossvein defects under certain conditions. As will be seen, there is quantitative phenotypic variation among genetically uniform individuals under apparently constant conditions. This variation is attributed to stochastic factors ("developmental noise"). Thus, the effect of adding a given number of *cve* alleles to an Oregon R genotype would be measured initially in terms of a mean phenotype.

A wild strain containing the amount of genetic variation generally seen in natural populations would be polymorphic at several *cve* loci. Thus, a genotype corresponding to the Oregon R base level genotype might have 4 more *cve* alleles than the minimum number possible in that wild strain. Part of a Poisson frequency distribution is sketched for a wild strain whose mean number of *cve* alleles on this scale was chosen to be 2.3.

B. *Incidence of Crossvein Defects in the Progenies of Wild Pairs*

The value 2.3 was chosen for consistency: given the scale of the model, it fits some general observations. A pair of flies would average $2 \times 2.3 = 4.6$ *cve* alleles between them, and about 10% of all pairs would have at least 8, or 4 per individual. The offspring would also average 4 *cve* alleles per individual, and (in a Poisson distribution again) at $18°C$ about 6% would have 7, 3% would have 8, and 1% would have 9 *cve* alleles. How many would exhibit crossvein defects?

Given the stochastic variation crudely estimated, about 10% of the flies with 7 *cve* alleles would be crossveinless at 18°C. Perhaps 20% with 8 and 50% with 9 *cve* alleles would show defects. This adds up to about 1% (the details are listed in Table I). Thus, 10% of all random pairs (those with at least 8 *cve* alleles between them) would have at least 1% *cve* offspring at 18°C. Obviously, this construct is just an illustration of the steps involved.

IV. THE SOURCES OF VARIATION

The parameters that relates the genotypes, phenotypes, and temperature is the underlying variable called crossvein making ability. It is closer to the genotype than to the phenotype in the sense that it is unaffected by temperature. Thus, genotypes can be described in terms of a standard parameter, crossvein making ability, which in turn predicts phenotypic distribution when the environment is specified.

A. *Environmental and Genetic Factors*

The sources of variation can now be reviewed. Under laboratory conditions, the environment is assumed to be effectively constant. Evidence for this assumption comes from the

TABLE I. *Proportion of the offspring[a] of pairs with 8 cve alleles in a population having a mean of 2.3 cve alleles per individual (4.6 cve alleles per pair giving 0.1 pairs in the population with at least 8 cve alleles).*

cve alleles (N)	proportion	probability of defects	product
7	0.06	0.1	0.006
8	0.03	0.2	0.006
9	0.01	0.5	0.005
Total proportion of crossveinless offspring			0.017

[a] post-F_1

observation that small differences in culture temperature and considerable differences in age of culture do not have substantial effects on mean crossvein ratings. The relevant genetic variation in the diverse natural populations sampled so far seems to be quite uniform in degree, though not in the specific loci involved. The empirical rule of thumb is that about 10% of all wild-inseminated females produce F_2's in which at least 1% of the flies have small crossvein defects at $18°C$. The wild-inseminated females are regarded as pairs, although it is now known that nearly half the *D. melanogaster* females trapped in nature are carrying sperm from more than one male (Milkman and Zeitler, 1973). Experience indicates that those isofemale lines with at least 1% crossveinless flies at the outset can be selected to yield true-breeding, polygenic *cve* strains with high mean ratings ($\geqslant 8$) at $25°C$. As Figure 2 indicates, this can be explained in terms of the segregation of a few *cve* alleles in the original population; the identification of parental "pairs" (inseminated females) carrying *cve* alleles at about 5 loci (some represented twice); and the subsequent fixation of these *cve* alleles. While the details of Figure 2 are offered only as an example, the important properties of the genetic variation described are presented as accurate. With this picture in mind, it is not difficult to envision the explicit analysis of this genetic complex in numerous populations and perhaps for the species as a whole.

B. *Stochastic Variation (Developmental Noise)*

The other important element in the initial phase of selection is stochastic variation. Were phenotypic variation limited to genetic causes, the 7-allele case illustrated in Figure 2 would be uniformly wild-type, and so would the 8-allele case. In fact, as we know, the genotype does not determine the phenotype absolutely, and the norm of reaction is simply the mean of a distribution that exists even in the absence of environmental variation. The present case is one in which the distribution of crossvein making ability associated with the 7-allele *cve* genotype extends past the first threshold of phenotypic expression at $18°C$. Thus, the difference between the 5-allele genotype and the 7-allele genotype is best described in terms of populations: at $18°C$, essentially no 5-allele individuals will have crossvein defects, while about 10% of the 7-allele individuals will. Thus, the 7-allele genotype carries a 10% chance of crossvein defects at $18°C$. The incidence of the crossveinless phenotype in the F_2 of a wild-inseminated female, thus, reflects the combinatorial distribution of *cve* alleles and the stochastic distribution of developmental events influencing crossvein making ability in those individuals whose norm of reaction is near the threshold of phenotypic expression.

V. POLYGENIC *CVE* STRAINS

A. *Selection*

In *D. melanogaster*, then, about one in ten wild isofemale lines will respond to 10-20 generations of selection with a true-breeding line of "crossveinless" flies. [The term *crossveinless* is used to indicate the absence of the entire crossvein or just part.] The combinations of alleles responsible for the defects are by no means the same every time. On the other hand, particular combinations of alleles do recur, and this observation opens a door to the study of crossveinlessness, of phenodeviants in general, and of even broader aspects of genetic and phenotypic variation.

Briefly to review the basic methodology and observations, dozens of *crossveinless* (*cve*) strains have been selected independently from the progenies of individual wild-inseminated *D. melanogaster* females. These strains have all proved to be polygenic, and all three major chromosomes are known to contain *cve* loci. A typical *cve* strain has the major portion of its *cve* activity on two chromosomes, and 3 to 5 loci are largely responsible for the deviant phenotype. Some mapping has been done; electrophoretic markers will make mapping a routine task from now on, as will be discussed later.

B. *Characterization of the Genotypes; the Identity Test*

While the various *cve* strains have been characterized in terms of phenotype (site and extent of defects), environmental cofactors (temperature during development), and genetic interactions (with simple recessives and with other polygenes), the critical property of a genotype is the loci involved. Thus, an extension of the classical allelism test was developed to determine whether two independent strains carried *cve* alleles at the same loci. This "identity test" consists, first, of crossing two *cve* strains. The F_1 is then examined: any marked change towards the normal phenotype indicates that the strains are different. Retention of the parental-level *cve* phenotype has two interpretations: genetic identity of the two parental strains (in terms of the *cve* loci involved) or merely positive interaction of their respective *cve* alleles, which might be at entirely different loci. It is thus necessary to determine whether a wild genotype can be reconstituted by recombination. Accordingly, the F_1 flies are allowed to inbreed for several generations, and then selection for the wild phenotype is instituted. Lines that do not respond to such selection are considered to come from identical *cve* parental strains.

Striking cases of "no response" have been observed (and confirmed) in a small proportion of tests (Milkman, 1965b, 1970b).

These tests were not conducted on a random sample of *cve* strains collected throughout the species range. Therefore, simple estimates of partial identity (one or more, but not all, loci shared in common) cannot be made. Nevertheless, the finding of complete identity between *cve* strains originating 3 miles and 6 months apart (Milkman, 1970b) suggests that the total number of loci in the *cve* library may be small enough to catalog. For example, if a truly random array of tests of 4-locus *cve* strains indicated identity in one case of every 16 (6.25%), a given locus would be shared in $(1/16)^{\frac{1}{4}}$ or 50% of all cases.

In the restricted theatres studied, more than 1/16 of all tests were positive. Furthermore, a modified test showed that often a *cve* strain carried additional unfixed *cve* alleles that were identical to those fixed in other strains (Milkman, 1970b). Thus, individual *cve* alleles are likely to recur frequently. This conclusion provides great incentive for the explicit demonstration of such recurrence.

In its original form, the identity test is effective only in distinguishing completely identical pairs of *cve* from less-than-completely identical (including totally dissimilar) pairs. This is because the crossveinless phenotype though itself quantitatively variable, is the visible part of a much larger range of variation in an underlying variable. Selection towards the normal phenotype soon becomes inefficient, and phenotypic variation has two possible individually sufficient causes: genetic and non-genetic. These are impossible to distinguish in the present case: numerous *cve* alleles may be segregating, and traits near a threshold are especially sensitive to environmental and stochastic variation (= developmental noise). Thus a new refinement was necessary to provide the required resolution.

C. *The Chromosome Identity Test*

The total identity test requires only one cross, namely, one between two *cve* strains, A and B. The chromosomal identity test requires several steps leading to the construction of

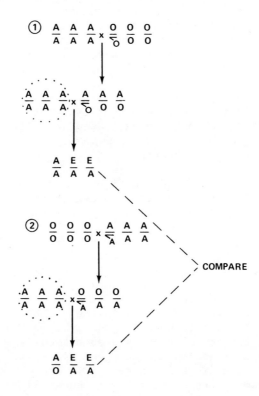

FIGURE 3. *A cross to determine whether a strain is carrying cve alleles on the X-chromosome.*

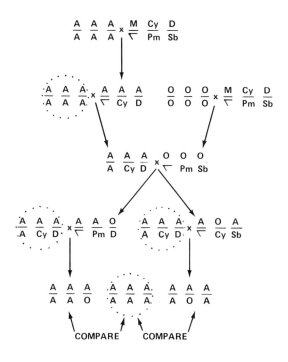

Direct visual comparisons, as illustrated in Figures 1 and 2, determine which cve chromosomes will be compared by identity testing.

FIGURE 4. Crosses to determine whether a strain is carrying cve alleles on the major autosomes.

a stock with one heterozygous (AB) major chromosome and two homozygous (AA) major chromosomes. The fourth chromosome is not controlled. If, after several generations of recombination and subsequent selection for the wild phenotype, no change is observed, the particular A and B chromosomes in question can be considered identical.

In practice, it is worth excluding major chromosomes with no *cve* effect, so that the crosses illustrated in Figure 3 and Figure 4 are performed first. These crosses will be recognized as the type generally used to localize genetic activity to particular chromosomes. Then, the procedure illustrated in

Figure 5 is employed for the appropriate autosomes. For the X-chromosome, the procedure illustrated in Figure 6 is one step shorter. In either case, the test centers on the effect of heterozygosity for a given chromosome in response to selection for normal crossveins.

The autosomal *crossveinless* loci can be mapped using electrophoretic markers, whose availability promises to revolutionize the localization of polygenes. (The X-chromosome has

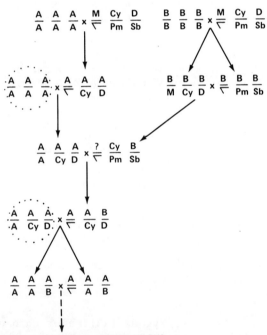

FIGURE 5. Crosses leading to the identity test for the major autosomes.

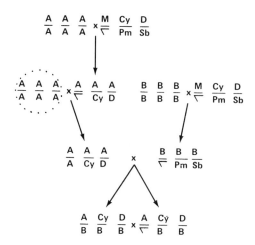

FIGURE 6. *Crosses leading to the identity test for the X-chromosome.*

never been a problem, because recessive visible markers can be used.) Unlike most of the small number of non-recessive morphological markers in *D. melanogaster*, electrophoretic variants are unambiguous, not associated with chromosomal aberrations, and widely distributed. Moreover, large scale analyses are now easy: one individual can genotype hundreds of flies at several loci in a day. Obviously each fly must be genotyped for all the electrophoretic loci used, and this is made possible by replicate sampling of the homogenate.

The procedure involves identifying the alleles at suitable loci in the *cve* chromosome to be mapped and choosing a *non-cve* stock with different electrophoretic alleles. Figure 7 illustrates the crosses.

VI. APPLICATIONS

While the most interesting chromosomes to map would be the recurrent one, any *cve* chromosome can be mapped in this way, and this leads to a consideration of the use of the data obtained. There are two applications of substantial promise.

First, a sizeable catalog of polymorphic genes can be obtained. Since loci are distinguishable by the identity test, high resolution mapping is not required to classify them. Until now, electrophoretic analysis has been the only detector

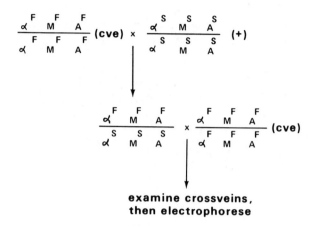

α − α-glycerophosphate dehydrogenase, II−20.5

M − malate dehydrogenase, II−41.2

A − alcohol dehydrogenase, II−50.1

FIGURE 7. *Example of mapping procedure using common electrophoretic variants.*

of a large class of polymorphic loci. The catalog of *cve* loci will be another step toward the enumeration of polymorphic loci in some species (in this case *D. melanogaster*). The genes need not be translated, incidently, to be detected, if indeed nontranslated genes play a role in morphological variation. In addition, cross comparisons between chromosomes associated with different deviant phenotypes can eventually be made. Clearly, the *cve* phenotype that serves as a paradigm here is but one of many deviant phenotypes.

The second application of the data goes more directly to the genetic basis of the crossveinless phenotype (or whatever deviant phenotype is studied). It is of interest to know just how many *cve* loci there are, or whether the number is too large to evaluate. [In the latter case, all is not lost: the catalog of polymorphic loci will gain correspondingly.] If the number is tractable, the distribution of the respective loci and their patterns of interaction will be of great interest: one of the major undescribed areas of biology is the distribution, within species, of alleles with specific phenotypic effects.

What hope is there for an explicit and essentially exhaustive enumeration of the loci associated with a phenodeviant? As Thoday and Thompson point out (1976), it does not take many polymorphic loci to produce a phenotypic distribution that is indistinguishable from continuous. My own observations indicate that a given *cve* strain differs from most flies in being homozygous for *cve* alleles at a very few loci. The fact that independent strains have occasionally proven to have identical sets of *cve* loci implies that individual loci must recur frequently, at least within the area sampled. Thus, unless the species range is a mosaic of areas in each of which certain *cve* alleles are unique, the hope of exhaustive enumeration is considerable. By the same token, a similar hope exists for other phenodeviants. And then we can begin to wonder about alleles associated with more than one familiar deviant phenotype.

VII. A POSTSCRIPT ON GENETIC ASSIMILATION

The major work on genetic assimilation, both conceptual and experimental, took place in the 1950's. Waddington's classic paper, "Genetic Assimilation of an Acquired Character", appeared in 1953, describing the acquisition of *crossveinless* strains by means of two-phase selection. First, the induction of crossvein defects by heat shock was the criterion of selection; later, as defects appeared spontaneously, they served as the criterion. This paradigm was studied further by K.G. Bateman, who also repeated the process with other phenotypes in *D. melanogaster*. Bateman's excellent studies concluded with a comparison of several selected *cve* strains; crosses among them were not analyzed beyond the F_1's, however. The title of Waddington's 1953 paper was intended (personal communication) to drive home the point that *all* characters are acquired in that the environment and pre-existing properties participate in shaping them. This viewpoint can be recognized in "The Strategy of the Genes" (1957) and especially in "Genetic Assimilation" which appeared in *Advances in Genetics* in 1960. From the latter view, it is clear that Waddington considered genetic assimilation as one view of the complex phenomenon of selection, rather than a discrete component with properties that are specific in some important way.

Since genetic assimilation, both in its experimental examples and its conceptual exposition, was a thinkpiece (indeed, a "thought-experiment" that was actually carried out); and since it represented an attempt to convey insight into the general processes of development and selection, rather than a distinct principle, it is not surprising that no important work on the subject has been done since 1960. The concept has continued to be useful in the purpose for which it was intended, and perhaps its most stimulating effect is one that it achieved at once (Simpson, 1953). It illuminates the coupling between learning and the evolutionary conversion of learned behavior into instinct. (Obviously not all instinct passes through a learning phase.)

There are three biological theatres in which the phenotypic consequences of the genotype are organized into several alternative coherent complexes. In a differentiated organism, genetically identical parts are structurally and functionally different. In a heterogeneous environment, genetically identical plants display different forms, which are functionally adaptive. In learning, identical twins can be brought up to speak different languages and pursue different trades.

What is general about these three situations is this: each elicits from a genotype a multiplicity of consequences for the scrutiny of natural selection and thus amplifies evolution. Simple viruses are much more rigid.

The causal equivalence, within limits, of genetic and environmental differences and the evolutionary impact of this limited equivalence, is the point of the principle of genetic assimilation.

VIII. A BRIEF REVIEW OF STUDIES ON THE POSTERIOR CROSSVEIN

Timoféeff-Ressovsky (1934) was the first to exploit the posterior crossvein as an indicator of extensive genetic variation in natural populations. His observations were extended to *Drosophila subobscura* by Gordon, Spurway, and Street (1939) and by Prevosti (1951, 1952). Clearly, the variants fit the description of isoalleles (Stern and Schaeffer, 1943). After Waddington's (1953) illuminating experiments, Bateman (1959) and Milkman (1960a) began to analyze the causality of crossvein defects. Mohler (1965, 1967) described additional genetic analysis later. Milkman (1960b) mapped some of the *cve* loci; subsequent localizations by various investigators are compiled in Milkman (1970a).

The role of temperature and the mechanism of high-temperature effects were explored in a series of papers (Milkman, 1961, 1962b, 1963, 1965a, 1966b; Milkman and Hille, 1966; Hille and Milkman, 1966) and subsequently reviewed (Milkman,

1967, 1970a). The distribution of *cve* alleles in natural and in laboratory populations was studied in another series of investigations (Milkman, 1960c, 1962a, 1964, 1966a; Boyer *et al.*, 1973). Other systems with similar general properties have been investigated in detail (see Thompson, 1975; Thompson and Thoday, 1975; and elsewhere in this volume).

These explicit and detailed observations of the interactions of genetic and environmental influences on a phenotype have led to the articulation of a particular approach to the mechanism of natural selection. This approach emphasizes the interaction of genes in causing a phenotype, and it emphasizes the phenotype as the criterion factor in selection (Milkman, 1967, 1978a, 1978b; Wills, 1978), as opposed to the analysis of selection on a locus-by-locus basis.

REFERENCES

Adamkewicz, S. Laura, and Milkman, Roger (1970). *Drosophila Information Service 45*, 98 (Abstract).
Bateman, K.G. (1959). *J. Genet. 56*, 443-474.
Boyer, Bradford J., Parris, Deborah L., and Milkman, Roger (1973). *Genetics 75*, 169-179.
Gordon, C., Spurway, H., and Street, P.A.R. (1939). *J. Genet. 38*, 37-90.
Hille, Bertil, and Milkman, Roger (1966). *Biol. Bull. 131*, 346-361.
Lerner, I. Michael (1954). "Genetic Homeostasis." Oliver and Boyd, Edinburgh.
Milkman, Roger (1960a). *Genetics 45*, 35-48.
Milkman, Roger (1960b). *Genetics 45*, 377-394.
Milkman, Roger (1960c). *Science 131*, 225-226.
Milkman, Roger (1961). *Genetics 46*, 25-38.
Milkman, Roger (1962a). *Genetics 47*, 261-272.
Milkman, Roger (1962b). *J. Gen. Physiol. 45*, 777-799.
Milkman, Roger (1963). *J. Gen. Physiol. 46*, 1151-1170.
Milkman, Roger (1964). *Genetics 50*, 625-632.
Milkman, Roger (1965a). *Genetics 51*, 87-96.
Milkman, Roger (1965b). *Genetics 51*, 789-799.
Milkman, Roger (1966a). *Genetics 53*, 863-874.
Milkman, Roger (1966b). *Genetics 54*, 793-803.
Milkman, Roger (1967a). *Genetics 55*, 493-495.
Milkman, Roger (1967b). *In* "Molecular Mechanisms of Temperature Adaptation" (C. Ladd Prosser, ed.). American Association for the Advancement of Science, Washington, D.C.
Milkman, Roger (1970a). *Adv. Genet. 15*, 55-114.

Milkman, Roger (1970b). *Genetics 65*, 289-303.
Milkman, Roger (1978a). *Genetics 88*, 391-403.
Milkman, Roger (1978b). In "Genetics, Ecology, and Evolution of Marine Organisms" (J. Beardmore and B. Battaglia, eds.) (In press).
Milkman, Roger, and Hille, Bertil (1966). *Biol. Bull. 131*, 331-345.
Milkman, Roger, and Zeitler, Rodney R. (1974). *Genetics 77*, 1191-1193.
Mohler, J.D. (1965). *Genetics 51*, 641-651.
Mohler, J.D. (1967). *Genetics 57*, 65-77.
Prevosti, Antonio (1951 and 1952). *Genetica Iberica 3*, 37-46, and *Genetica Iberica 4*, 95-128. (Abridged, combined translation of the two papers by Roger Milkman.)
Simpson, G.G. (1953). *Evolution 7*, 110.
Stern, C., and Schaeffer, E.W. (1943). *Proc. Natl. Acad. Sci. U.S.A. 29*, 361-367.
Thoday, J.M., and Thompson, J.N. (1976). *Genetica 46*, 335-344.
Thompson, J.N. (1975). *Heredity 35*, 401-406.
Thompson, J.N., and Thoday, J.M. (1975). *Genet. Res., Camb. 26*, 149-162.
Timoféeff-Ressovsky, N.W. (1934). *Nachrichten (Biologie) von der Gesellschaft der Wissenschaften zu Göttingen. Mathematischphysikalische Klasse. Neue Folge. Fachgruppe VI. 1*, 53-104. (Translation by Roger Milkman).
Waddington, C.H. (1953). *Evolution 7*, 118-126.
Waddington, C.H. (1957). "The Strategy of the Genes." Macmillan, New York.
Waddington, C.H. (1961). *Adv. Genet. 10*, 257-293.
Wills, C. (1978). *Genetics 89*, 403-417.

*Analysis of Individual
Gene Effects*

POLYGENIC MUTATION

Terumi Mukai

Department of Biology
Kyushu University
Fukuoka, Japan

I. INTRODUCTION

The astonishingly rapid development of molecular genetics has supplied us with abundant information of major genes, but little is known about polygenes, which are supposed to be very important for evolution. The definition of polygenes was operational. They are "the genes having effects similar and supplementary to one another, and small in comparison with the total variation", and the system of polygenes controlling a specific continuous variable is called a polygenic system (Darlington and Mather, 1949). According to this definition, pleiotropic effects of mutations of major genes might be included among polygenic mutations.

In *Drosophila melanogaster*, studies on polygenic mutations mainly involved the variables viability and sternopleural and abdominal bristle numbers. In the present chapter, the estimation of the mutation rates of polygenes affecting these two kinds of characters will be discussed. There is a large difference between the two in the methods of estimation of mutation rates since the directions of the average effects of mutations affecting them are different.

II. METHODS AND EXAMPLES FOR THE ESTIMATION OF POLYGENIC MUTATION RATES

In the expression of the rate of polygenic mutation, the unit for the major genic mutation rate, or the rate per locus, can not be used since we do not know the number of loci involved.

There are two ways of expressing the mutation rate of polygenes. The first is to give the number of mutant genes per chromosome or per genome per generation or per R of X-rays or γ-rays. The minimum number of mutant polygenes can be estimated only when population means and variances are changed by mutation. The second way, which was first proposed by Clayton and Robertson (1955), is to express the rate of polygenic mutations in terms of the increment of genetic variance in the population per R of X-rays or γ-rays or per generation. This increment can be compared with the genetic variance in the equilibrium population, and is meaningful from the evolutionary view-point.

A. *Minimum Mutation Rate and Maximum Average Effect of Individual Mutations*

This attempt was made first by Bateman (1959) and the method was further advanced by Mukai (1964) and Mukai *et al.* (1972). A mathematical model is set up under the following assumptions:

(1) polygenic mutations are distributed on a chromosome or genome according to a Poisson distribution,

(2) individual effects of polygenic mutations are s_i (i = 1-n). The average effect is $\bar{s}$ and the variance of individual mutations is σ_s^2,

(3) polygenic mutations occur at an average frequency of p in one generation on a chromosome or in a genome,

(4) additivity holds true among loci and the mean of the polygenic character changes at a rate of M and the genetic variance increases at a rate of V per generation.

It is necessary to estimate M and V experimentally. The expected values of M and V are calculated as follows:

$$\bar{s}p = M \tag{1}$$

$$(\bar{s}^2 + \sigma_s^2)p = V \tag{2}$$

From equations (1) and (2), the following equations can be obtained:

$$p = (M^2/V)(1 + K) \tag{3}$$

$$\bar{s} = (V/M)[1/(1 + K)] \tag{4}$$

and

$$\sigma_s^2 = (V/2M)^2 \left[1 - \{(1-K)/(1+K)\}^2\right] \tag{5}$$

where K is $\sigma_s^2/\bar{s}^2$ or the square of the coefficient of variation of s values.

From equations (3), (4), and (5), the following equations can be obtained:

$$p \geqq \frac{M^2}{V} \tag{6}$$

$$\bar{s} \leqq \frac{V}{M} \tag{7}$$

$$\sigma_s^2 \leqq (\frac{V}{2M})^2 \tag{8}$$

From these equations, the minimum of polygenic mutation rate per chromosome or per genome, the maximum of the average effect of individual polygenic mutations and the maximum variance of individual effects can be estimated.

An example of estimation of the rate of polygenic mutations affecting viability and its maximum average effect can be taken from the works of Mukai (1964) and Mukai et al. (1972). A single male of D. melanogaster carrying normal second chromosomes was crossed to $In(2LR)SM1/In(2LR)bw^{V1}$ female flies. $In(2LR)SM1$ stands for a multiple inversion marked by the dominant gene Cy, and $In(2LR)bw^{V1}$ is another multiple inversion carrying a dominant marker Pm. The former is abbreviated Cy and the latter Pm. From the offspring, a single male Pm/+ (or Cy/+) was picked out. This was crossed to as many Cy/Pm females as possible. From the offspring, n $Pm/+_i$ (i = 1 - n) males were sampled. These males were individually crossed to 5 Cy/Pm females and n chromosome lines were established. In each line, $+_i$ chromosomes were maintained by the cross Cy/Pm (5 females) × $Pm/+_i$ (1 male). These mating schemes are shown in Figure 1. Following this method, mutations can be accumulated at a minimum pressure of natural selection, because every mutation established in the line is subject to natural selection only for one generation in heterozygous condition. At certain generations the homozygous viability of the second chromosome in each line ($+_i$) was estimated by the Cy method of Wallace (1956), namely that several crosses of five $Cy/+_i$ × five $Cy/+_i$ were done. In the offspring, two types of flies, $Cy/+_i$ and $+_i/+_i$ flies segregate at a ratio of a:b. The expectation is 2:1. The viability of $+_i/+_i$ can be expressed by 2b/(a + 1), 1 in the denominator

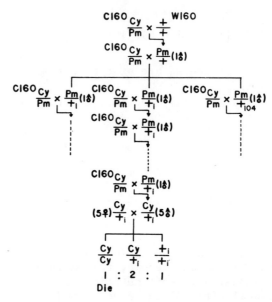

FIGURE 1. *Mating scheme for accumulation of mutations (from Mukai, 1964).*

being Haldane's (1956) correction. The estimations of viabilities were done at generations 10, 15, 20, and 25 in the first experiment (Mukai, 1964) and at generations 10, 20, 30, and 40 in the second experiment (Mukai et al., 1972). In the first experiment a single original second chromosome was employed, but in the second experiment three were used. From the distribution of viabilities of all lines, it was found that mutations were classified into three categories: lethal, semilethal, and mildly deleterious mutations. It is the last category that was of interest in this work, and the lines carrying lethal and semi-lethal mutations were excluded. The decrease in mean viability and the increase in genetic variance with generation in the second experiment are graphically shown in Figure 2. For the estimation of genetic variance, one-way analysis of variance was applied. From Figure 2, it is seen that mean viability decreased and the genetic variance increased, both approximately linearly with generation. Thus, the rate of decrease in the mean and of increase in the variance were estimated applying a linear regression. The results were $\hat{M} = 0.00403$ and $\hat{V} = 0.000094$. The same calculation was done for the result in the first experiment. The results were $\hat{M} = 0.00378$ and $\hat{V} = 0.000101$. Applying these estimates to equations (6), (7), and (8), p, $\bar{s}$, and σ_s^2 were estimated and

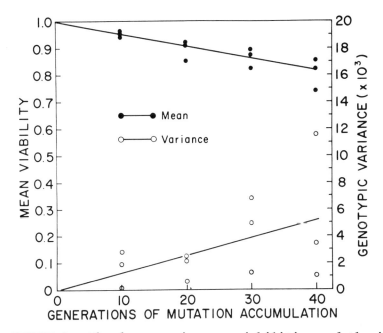

FIGURE 2. The decrease in mean viabilities and the increase in genetic variances for viability with accumulation of polygenic mutations (from Mukai et al., 1972).

all the results are given in Table I. From this table, it can be seen that the rate of polygenic mutations affecting viability is approximately 20 times higher than the rate of recessive lethal mutations, and that individual effects of mutant polygenes are of the order of 2-3% of the normal viability.

B. *The Rate of Increase in Genetic Variance (Analysis of Variance)*

1. Factorial Experiment (Partial Diallel Cross). If many inbred lines (or equivalents) are available from a population where mutations are occurring, the following factorial experiment [or North Carolina Design II (Comstock and Robinson, 1952) or partial diallel cross] can be used for the estimation of genetic variance. At first we choose $(a + b)$ chromosome lines at random. The a lines will be used for maternal parents and the remainder (b lines) will be used for paternal parents and an $a \times b$ factorial experiment will be conducted

TABLE I. Results of the experiments for estimating spontaneous mutation rates of polygenes affecting viability.

	Mukai (1964)	Mukai et al. (1972)
Rate of mean decrease[a]	0.00378	0.00403
Rate of variance increase[a]	0.000101	0.000094
$\hat{\bar{s}} <$	0.027	0.023
$\hat{\sigma}_s \leq$	0.013	0.012
$\hat{p}$ $a \geq$	0.141	0.172
Lethal mutation rate[a]	0.0063	0.0060

[a] per second chromosome per generation

with n simultaneous replications.

The sums of squares due to female parents (SS_F), male parents (SS_M), their interactions ($SS_{F \times M}$) and errors (SS_E) can be estimated following the general procedure described in the text book of statistics. The mean squares corresponding to the sums of squares and their expected values are described in Table II, where σ_M^2, σ_P^2, and $\sigma_{M \times P}^2$ are variances among the contributions of maternal parents, among the contributions of paternal parents, and the variance of the contributions of interaction between paternal and maternal parents, respectively. σ_E^2 stands for the environmental variance.

On the basis of the results described in Table II, σ_M^2, σ_P^2, and $\sigma_{M \times P}^2$ can be estimated as follows:

TABLE II. Analysis of variance for factorial experiments.

Source of variation	D.F.	Mean square	Expected M.S.
Maternal	$a-1$	MS_M	$\sigma_E^2 + n\sigma_{M \times P}^2 + bn\sigma_M^2$
Paternal	$b-1$	MS_P	$\sigma_E^2 + n\sigma_{M \times P}^2 + an\sigma_P^2$
M×P Interaction	$(a-1)(b-1)$	$MS_{M \times P}$	$\sigma_E^2 + n\sigma_{M \times P}^2$
Error	$ab(n-1)$	MS_E	σ_E^2

$$\hat{\sigma}_M^2 = \frac{MS_M - MS_{M\times P}}{bn} \qquad (9)$$

$$\hat{\sigma}_P^2 = \frac{MS_P - MS_{M\times P}}{an} \qquad (10)$$

$$\hat{\sigma}_{M\times P}^2 = \frac{MS_{M\times P} - MS_E}{n} \qquad (11)$$

When inbred lines were used as parents, σ_M^2, σ_P^2, and $\sigma_{M\times P}^2$ can be related to additive variance (σ_A^2) and dominance variance (σ_D^2) as follows (c.f., Mukai et al., 1974):

$$\sigma_A^2 = 2\sigma_M^2 = 2\sigma_P^2 = \sigma_M^2 + \sigma_P^2 \qquad (12)$$

$$\sigma_D^2 = \sigma_{M\times P}^2 \qquad (13)$$

The increment of genetic variance due to spontaneous mutations (or artificially induced mutations) ($\Delta\sigma_G^2$) can be estimated as follows:

$$\Delta\hat{\sigma}^2 = (\hat{\sigma}_A^2 + \hat{\sigma}_D^2)/\text{generation number} \qquad (14)$$

In the case of radiation-induced mutations, $\hat{\sigma}_A^2 + \hat{\sigma}_D^2$ should be divided by the dose of radiation.

Mukai (1977) conducted a factorial experiment using experimental materials in which spontaneous mutations were accumulated for 160 generations (see section I.A.). Since mutations were accumulated on 150 chromosome lines which originated from 3 independent chromosomes (50 lines from each stem chromosome), nine of 8 × 8 factorial experiments ($Cy/+_i \times Cy/+_j$, i = 1-8, j = 9-16) were conducted with four simultaneous replications, where the genetic backgrounds of all crosses were arranged to be homogeneously heterozygous. The viabilities were estimated by the Cy method. The components of genetic variance that were estimated taking into account the incomplete dominance of the Cy chromosome (see Mukai et al., 1974) were: $\hat{\sigma}_A^2 = 0.0209\pm0.0037$ and $\hat{\sigma}_D^2 = 0.000239\pm0.000495$. Thus, the rate of increase in genetic variance ($\Delta\hat{\sigma}_G^2$) is:

$$\Delta\hat{\sigma}_G^2 = 0.000154/\text{generation}.$$

The $\hat{\sigma}_A^2$ and $\hat{\sigma}_D^2$ in the Raleigh, N.C., population were: $\hat{\sigma}_A^2 = 0.00960\pm0.00253$ and $\hat{\sigma}_D^2 = 0.00115\pm0.00046$. Thus, about 70 generations are necessary to accumulate the genetic variance of viability that is found in equilibrium populations.

2. *Nested Design (or Hierarchical Classification)*. When inbred lines are not available, use has to be made of a sample from a random mating population. A sample of individuals is taken from a random mating population and irradiated or chemically treated. A total of a males are taken and they are individually mated to b females. From the offspring of each family, n individuals are randomly sampled and their quantitative characters are measured. The analysis of variance technique (hierarchical classification or nested design) can be applied and the total sum of squares can be partitioned into Male parents, Female parents within Male parents, and Individuals within full-sib families. The analysis of variance table is given in Table III. From this table, the σ_M^2, $\sigma_{F(M)}^2$, and σ_I^2 which are population variances of male contributions, female contributions within male families, and of the effects of segregation within full-sib families plus the effects of environmental variations, can be estimated as in the factorial experiment. The additive and dominance variances can be estimated from the following relationship:

$$\sigma_A^2 = 4\sigma_M^2 \tag{15}$$

$$\sigma_D^2 = 4(\sigma_{F(M)}^2 - \sigma_M^2) \tag{16}$$

The variance due to environmental effects (σ_E^2) can be estimated as follows:

$$\hat{\sigma}_E^2 = \hat{\sigma}_I^2 + \hat{\sigma}_M^2 - 3\hat{\sigma}_{F(M)}^2 \tag{17}$$

TABLE III. *Analysis of variance for nested design.*

Source of variation	D.F.	Mean square	Expected M.S.
Paternal	$a-1$	MS_P	$\sigma_I^2 + n\sigma_{M(P)}^2 + bn\sigma_P^2$
Maternal within Paternal	$a(b-1)$	$MS_{M(P)}$	$\sigma_I^2 + n\sigma_{M(P)}^2$
Individual	$ab(n-1)$	MS_I	σ_I^2

Tobari and Nei (1965) employed this method. They irradiated a sample of *D. melanogaster* taken from a random mating population with 1,000R of X-rays. After two generations of random mating, they took a sample of 50 males and 200 females and mated each individual male to four females. Sternopleural and abdominal bristle numbers were counted for 10 females from each full-sib family. The control experiments were conducted without irradiation. For sternopleural bristles, a second experiment was conducted with 102 males and 302 females. Since they found that, by chance, the additive variance was decreased but that the dominance variance was increased with irradiation in sternopleural bristle number, the two estimates were pooled and the increments of genotypic variances were estimated for both characters.

The rates of increase in genetic variance were calculated as follows:

Abdominal bristle number: (X-rayed) − (control)
 = 9.4974 − 8.4836 = 1.0138/1000R (or 1.01×10^{-3}/R)

Sternopleural bristle number: (X-rayed) − (control)
 (The first experiment) = 2.3320 − 1.4398
  = 0.8922/1000R
 (The second experiment) = 1.9655 − 1.8616
  = 0.1039/1000R

The average = 0.50×10^{-3}/R

C. *The Rate of Increase in Genetic Variance (Selection Experiments)*

If the genetic variance consists of additive genetic variance alone (no dominance), it can be estimated by selection experiments. Let us suppose that we begin the selection experiment using an inbred line that has been maintained by brother-sister mating for many generations. Genetic variance at the starting generation $[\sigma_g^2(0)]$ can be expressed by the following formula:

$$\lambda \sigma_g^2(0) = \sigma_{gm}^2 \qquad (18)$$

where λ is the rate of change in heterozygosity and σ_{gm}^2 is the genetic variance caused by spontaneous mutations in one generation. Thus, we obtain the following relationship ($\lambda = 0.191$):

$$\sigma_g^2(0) = \frac{1}{\lambda} \sigma_{gm}^2 = 5.2 \sigma_{gm}^2 \qquad (19)$$

At generation t of the selection experiment, the following relationship may hold:

$$\sigma_g^2(t) \stackrel{\sim}{=} (1 - \frac{1}{2N})\sigma_g^2(t-1) + \sigma_{gm}^2 \qquad (20)$$

where N is the effective size of the population. If $\sigma_g^2(0)$ is zero, the following formula can be obtained from formula (21) (Clayton and Robertson, 1964):

$$\sigma_g^2(t) \stackrel{\sim}{=} 2N\sigma_{gm}^2 (1 - e^{-\frac{t}{2N}}) \qquad (21)$$

From a selection experiment, the genetic variance at generation t can be estimated using the following relationship, where ΔG is the genetic gain:

$$\Delta G = \frac{\sigma_g^2}{\sigma_P^2} \cdot \bar{i}\sigma_P \qquad (22)$$

therefore, $\quad \sigma_g^2 = \Delta G \cdot \sigma_P / \bar{i} \qquad (23)$

where σ_P^2 is the phenotypic variance, $\bar{i}\sigma_P$ is the selection differential, where $\bar{i}$, depending on the degree of selection, can be obtained from tables (Fisher and Yates, 1938).

Clayton and Robertson (1955) conducted a selection experiment for the number of bristles on the 4th and 5th abdominal segments in *D. melanogaster* as the control of a radiation experiment. They used an inbred line (Oregon) maintained for many generations by brother-sister mating. The selection procedure was as follows: 25 females and 25 males were measured and the extreme 10 of each sex were selected as parents for the next 17 generations.

In the equilibrium population of brother-sister mating, $\sigma_g^2(0) \stackrel{\sim}{=} 5.2\sigma_{gm}^2$. From formula (21), $\sigma_g^2(14) \stackrel{\sim}{=} 15\sigma_{gm}^2$. Generation number 14 is the average of 11-17. Therefore, the genetic variance at generation 14 due to the new spontaneous mutations after the initiation of the experiment $[\sigma_g^2(14)]$ is:

$$\sigma_g^2(14) \stackrel{\sim}{=} 15\sigma_{gm}^2 - 5\sigma_{gm}^2 = 10\sigma_{gm}^2. \qquad (24)$$

From the actual experiment ΔG at generation 14 was estimated to be 0.027 (actually, this value is the average response per generation over about 14 generations). Thus, from formula (23), $\sigma_g^2(14)$ can be estimated to be:

$$\hat{\sigma}^2_g = \Delta G \cdot \sigma_p / \bar{i} = 0.027 \times 2.1/0.94$$

$$= 0.060 \quad (25)$$

where $\bar{i}$ was 0.94 (corresponding to 10 selected out of 25) and $\hat{\sigma}_p = 2.1$ (from the data). From (24) and (25), $\hat{\sigma}^2_{gm}$ becomes 0.006.

III. GENERAL REVIEW OF THE EXPERIMENTAL RESULTS

A. *Viability*

1. Spontaneous Mutations. Spontaneous polygenic mutations affecting viability were first examined by Durrant and Mather (1954). They sampled second chromosomes from an inbred Oregon line of *D. melanogaster* maintained for more than 300 generations of continuous brother-sister mating, and tested these for heritable variations affecting viability. They found significant heritable variation.

By contrast, Paxman's (1957) experimental result might be mentioned. He accumulated viability mutations on seven second chromosomes that originated from an inbred line (Oregon), and found neither a change of overall mean nor any difference between lines.

As described above, Mukai (1964) and Mukai *et al.* (1972) have accumulated mutations affecting viability, and found that spontaneous mutation rate of viability polygenes is approximately 20 times higher than that of recessive lethals per chromosome. This finding was confirmed by Ohnishi (1977).

Mukai and Yamaguchi (1974) and Watanabe *et al.* (1976) estimated the homozygous load of the second and third chromosomes of *D. melanogaster* which were collected from the Raleigh, N.C., population. Assuming that polygenic variation with respect to viability was maintained by a balance between mutation and selection pressures, they estimated the rate of spontaneous polygenic mutations affecting viability on the basis of homozygous loads (Greenberg and Crow, 1960) and various estimates of population-genetic parameters in the population such as degrees of dominance of different kinds of mutations. The results have supported the above finding.

Comparisons were made between the genetic variation found in approximately equilibrium *D. melanogaster* populations and that created by spontaneous mutations with respect to homozygous load and genetic variance for viability. Homozygous loads were estimated for the Raleigh, N.C., population and for the population carrying only newly-arisen mutations which were accumulated at a minimum pressure of natural selection (Mukai and Yamazaki, 1968; Mukai et al., 1972; Mukai and Yamaguchi, 1974). Homozygous lethal load (L) and detrimental load (D), measured as a deviation from the average heterozygote were estimated as follows:

$$D = \ln(A/C), \text{ and } L = \ln(C/B)$$

where A is the average viability of heterozygotes, B is the average viability of all homozygotes, and C is the average viability of non-lethal homozygotes (Greenberg and Crow, 1960). The genetic variances were estimated for these two populations (Mukai et al., 1974; Mukai, 1977). These results are shown in Table IV. From this table, it is clear that the two results are mutually consistent and that the level of genetic variation with respect to viability (both lethal genes and viability polygenes) to be found in approximately equilibrium populations can be reached by accumulating spontaneous mutations for less than 100 generations.

TABLE IV. *Comparison between the genetic variability with respect to viability in an approximately equilibrium population and that produced by spontaneous mutations (the second chromosome). (From Mukai et al., 1972; Mukai and Yamaguchi, 1974; and Mukai, 1977).*

	New mutations	Mutations in equilibrium population	Ratio
HOMOZYGOUS LOAD[a]			
Viability polygenes	0.0040/ generation	0.30 (0.47)[b]	74 (118)[b]
Lethal mutations	0.0060/ generation	0.50	79
GENETIC VARIANCE			
Total	0.0001321	0.01075	81
Additive	0.0001306	0.00960	74

[a] *The unit is lethal equivalent.*

[b] *The figures in parentheses are adjusted value (the standard is the optimum genotype in the population).*

2. Induced Mutations. Apparently mutually contradictory results have been reported with respect to the rate of radiation-induced polygenic mutations affecting viability. The first attempts to obtain the distribution and frequency of detrimental mutations induced by radiation were done by Timoféeff-Ressovsky (1935) and Kerkis (1935). According to them, detrimental mutations (mutant viability polygenes must be included) were induced two to three times as frequently as lethals in *D. melanogaster*. Later, Käfer (1952) tested 500 X and second chromosomes with respect to viability after irradiation with X-rays at doses of 2.5-7.5KR, and found statistically significant detrimentals at a frequency of 75% of complete lethals. Bonnier and Jonsson (1957) obtained a similar result.

Friedman (1964) estimated the detrimental load to lethal load ratio (D : L ratio) after irradiating X chromosomes with 1000R, 3500R, and 7000R. The result turned out to be 0.125 and supported the findings obtained by Käfer and by Bonnier and Jonsson. Simmons (1976) also supported Friedman.

Consideration of the ratio of homozygously deleterious mutations (at least a 10% reduction in normal viability) to recessive lethal mutations resulting from radiation in *D. melanogaster* (Muller, 1934; Kerkis, 1938; Timoféeff-Ressovsky, 1935) led Muller (1950) to predict that the rate of mutations, whose viability reductions are less than 10% (therefore, they might be called polygenic mutations), may be much more frequent than all the more markedly detrimental mutations taken together. This appears to have been substantiated by Wallace (1957, 1958, 1959), Burdick and Mukai (1958), Bateman (1959), and Mukai *et al.* (1966).

Wallace, Burdick and Mukai, and Mukai *et al.* tested the heterozygous effects of mutations induced in the second chromosomes with low doses of X-ray irradiation, and found that on the average they were beneficial (1.5-3.0% increase). Wallace and Mukai *et al.* irradiated flies with 500R of X-rays. These results cannot be explained unless the radiation-induced mutation rate were high. Indeed, the expected frequency of recessive lethal mutations is only 0.027/second chromosome/500R. Maruyama and Crow (1975) considered the positive heterozygous effect of radiation-induced mutations as due to some unknown causes other than mutations, although they themselves obtained the same result. However, this type of overdominance was also found for spontaneous (Mukai *et al.*, 1964; Mukai and Yamazaki, 1968) as well as for EMS-induced mutations (Mukai, 1970).

Bateman (1959) is the first person to attempt to estimate the radiation-induced polygenic mutation rate affecting viability per chromosome using an advanced method described in section I,A. He employed the data obtained by Lüning and Jonsson (1957) who estimated the hemizygous effects of mutations induced in the X chromosomes of *D. melanogaster* with 2160R and 3240R of X-rays. According to Bateman (1959), the minimum numbers of mutations per X chromosome range from 1.35 to 18.53 with the irradiation of 2160R and 0.04 with 3240R. Although there is a large variation in the estimates of mutation rates, these results definitely indicate that the rate of radiation-induced mutations affecting viability is high on a chromosome basis.

Using the data of Mukai *et al.* (1966) described above, the number of mutant genes (p), the average heterozygous effect of individual mutant genes (s), and the variance of the individual effects (σ_s^2) were estimated by them using equations (6), (7), and (8). The results are: $0.786 \leqq p$, $\bar{s} \leqq 0.0384$, and $0 \leqq \sigma_s^2 \leqq 0.000369$, where $\hat{M} = 0.0302$ and $\hat{V} = 0.001161$. Although the reliability of these estimates might not be high because the estimation was conducted on the basis of the ratio of two estimates probably having large coefficients of variation, the results together with those of Bateman (1959) indicate that the rate of radiation-induced polygenic mutations affecting viability is extremely high. The above rate is 26.7 times larger than the recessive lethal mutation rate (Mukai *et al.*, 1966), which is almost equal to the ratio in spontaneous mutations.

For the sake of reference, the D : L ratio of Greenberg and Crow (1960) was surveyed. The results are shown in Table V. The ratios for induced mutations including chemically induced mutations are much smaller than those for spontaneous mutations. Thus, if Bateman (1959) and Mukai *et al.*'s (1966) results are generally accepted, they suggest that radiation induces many polygenic mutations with very small effects which do not significantly contribute to the detrimental load. This conforms to the expectation of Muller (1950) as described above, but further experimental evidence is required.

B. *Bristle Characters*

Since Serebrovsky (1935) and Rokizky (1936) studied the effects of X-rays on the sternopleural bristle number of *D. melanogaster*, several investigators have studied polygenic mutations affecting bristle characters of this species. Information concerning spontaneous mutation is more limited.

TABLE V. Detrimental load to lethal load ratio in spontaneous, X-ray induced, and chemically induced mutations.

	Detrimental load (D)	Lethal load (L)	Total load (T)	Ratio (D/L)	Literature[a]
Spontaneous mutation	0.180	0.184	0.364	0.978	1
	0.225	0.236	0.461	0.953	2
X-ray induced mutation	0.106	0.149	0.254	0.711	3
	0.068	0.234	0.302	0.288	4
	0.037	0.172_5	0.209	0.215	5
	3.46×10^{-6}/R	2.76×10^{-5}/R	3.11×10^{-5}/R	0.125	6
EMS-induced mutations	0.240	0.535	0.775	0.449	7[b]
	0.0790	0.2779	0.3569	0.28	8[c]

[a] 1, Mukai and Yamazaki, 1968; 2, Mukai et al., 1972; 3, Timoféeff-Ressovsky, 1935; 4, Käfer, 1952; 5, Bonnier and Jonsson, 1955; (3, 4, and 5 are from Greenberg and Crow, 1960); 6, Friedman, 1964; 7, Mukai, 1970; 8, Ohnishi, 1977.
[b] 2.5×10^{-2} mol of EMS was given.
[c] 2.5×10^{-3} mol of EMS was given.

TABLE VI. The rate of increase in genetic variance of sternopleural and abdominal bristle numbers in D. melanogaster (whole genome and heterozygous)

Author	Abdominal bristle number	Sternopleural bristle number
(1) Spontaneous mutation (per generation)		
Mather and Wigan (1942)	0.007	0.002
Durrant and Mather (1954)	0.014	0.005
Clayton and Robertson (1955)	0.006	---
Paxman (1957)[a]	0.0013	0.0007
(2) Radiation-induced mutation (per R)		
Clayton and Robertson (1955)	3.3×10^{-5}	---
Scossiroli and Scossiroli (1959)	---	10.7×10^{-5}
Yamada (1961)	38.7×10^{-5}	13.9×10^{-5}
Tobari and Nei (1964)	101×10^{-5}	50×10^{-5}
Clayton and Robertson (1964)	1×10^{-5}	---
Kitagawa (1967)	29.6×10^{-5}	---

[a] The present author's calculation from Paxman (1957).

1. *Spontaneous Mutations*. Mather and Wigan (1942) and Clayton and Robertson (1955) conducted experiments using the Oregon inbred lines which had been maintained by brother-sister mating for 78 generations and 28 generations, respectively. Using the methods for the experiment and the analysis described above, the genetic variances for abdominal bristle number arising by spontaneous mutations were estimated to be 0.007 and 0.006 (upper limit) per generation, respectively. The rate of increase in genetic variance for sternopleural bristle number for the Mather and Wigan (1942) experiment was 0.002 per generation (Clayton and Robertson, 1955).

Durrant and Mather (1954) estimated the genetic variance existing in an inbred line (Oregon) maintained by brother-sister mating for over 300 generations. They used the diallel cross method, testing 10 second chromosomes, in a way similar to that described above (section I.B.1.). The variances due to chromosome differences were estimated to be 0.0203 and 0.0562 for sternopleural and abdominal bristle number, respectively. The resulting estimates of σ^2_{gm} or the genetic variance produced by spontaneous mutations per generation for sternopleural and abdominal bristle numbers, respectively, are presented in Table VI.

The genetic variance in approximately equilibrium populations is about 5 for abdominal bristle number and about 2 for sternopleural bristle number. Thus, accumulation of spontaneous mutations for about 800 generations is necessary to obtain the genetic variance of the equilibrium population. This is about 10 times larger than viability mutations, and suggests that some mechanisms other than the balance between mutation and selection pressures is operating for the maintenance of genetic variation of these characters.

2. *Radiation-induced Mutations*. The main interest of the investigators who studied the radiation-induced mutations concerned the role of such mutations in responses to artificial selection. Unfortunately, rather contradictory results have been obtained. Scossiroli (1954) used two lines that Professor Mather had selected for the number of sternopleural bristles until response ceased. Scossiroli's experimental lines were irradiated in each cycle of two generations with 3000R of X-rays, and family selection was conducted in every second generation for high bristle number, applying a selection pressure of 16.78% giving a selection differential of 1.45 standard deviations. After 17 cycles of selection, bristle number had risen from 26 to 46 on the average. However, there remained one doubt that these responses might have been caused by radiation-induced recombination, namely that linkages of + and - genes were broken by X-rays and newly

formed *+ +* chromosomes might have been picked out by selection. This doubt was cleared by Scossiroli and Scossiroli (1959), who conducted the selection experiment for sternopleural bristle numbers of *D. melanogaster* with both isogenic lines (Oregon) and the cross between two isogenic lines (Oregon × Samarkand). Flies were irradiated every generation or every other generation with 3000R of X-rays and there were clear responses to selection. Kitagawa (1966) obtained similar results using abdominal bristle number and irradiating flies with 1500R of X-rays in every generation. According to their calculation (Yamada and Kitagawa, 1961; Kitagawa, 1966), the rates of increase in genetic variance were 10.73×10^{-5}/R for the Scossiroli and Scossiroli (1959) results for sternopleural bristle number and 29.6×10^{-5}/R for Kitagawa's (1966) results for abdominal bristle number.

As described above, Tobari and Nei (1964) showed that the rates of increase in genetic variance were 101×10^{-5}/R and 50×10^{-5}/R for abdominal and sternopleural bristle numbers, respectively, much higher than the above.

Using the marked inversion chromosomes (Cy and Ubx^{130}) efficiently, Yamada (1961) directly estimated the rate of increase in variance for both the sternopleural and abdominal bristle numbers after irradiating one set of second and third chromosomes per individual with 250-2000R of γ rays. After that the rates were converted to a genome basis (diploid). The results were 38.7×10^{-5}/R for abdominal and 13.9×10^{-5}/R for sternopleural bristle numbers.

By contrast, Clayton and Robertson (1955) found that 1800R/generation of X-rays gave no increase in the response to selection for abdominal bristle number. Their selection procedure was described above (section I, C). After continuation of selection for 17 generations, they only achieved a difference of 3.57 bristles between the lines selected upwards and those selected downwards. In a second experiment (Clayton and Robertson, 1964), seven plateaued populations as well as three different inbred lines were used. Although some of the irradiated populations showed a greater response to selection than the control, the increment of genetic variance was only approximately 1×10^{-5}/R. This value and that in their first experiment (3.33×10^{-5}/R) are much smaller than those for the other experiments described above. The rates of variance increases are tabulated in Table VI. At present we do not know the reason for these differences. It may, however, be that the frequency of some new mutations can sometimes be increased, despite the pressure of natural selection, by artificial selection but others are eliminated from the population by natural selection. If so, the rate of increase in variance resulting from mutation would be larger than that estimated in a selection experiment.

IV. DISCUSSION

From the summary of the studies on mutations of polygenes, it is clear that in every generation a certain amount of genetic variability is supplied to the population by mutation. The variation of viability polygenes is primarily maintained by the balance between mutation and selection pressures. This can be understood from the comparison with lethal genes which are known to be deleterious in heterozygotes (see Table IV). However, some fraction appears to be maintained by some form of balancing selection (Mukai and Yamaguchi, 1974). The mechanism for the maintenance of genetic variation of bristle characters is different. This can be seen from the difference in the number of generations required for the accumulation of the same amount of genetic variance as that in the equilibrium population (see section II, B, 1). The optimum model with mutations (Kimura, 1965) or overdominance with respect to fitness, but additive with respect to the character itself (Robertson, 1956), may be the mechanism by which the variation of the bristle characters is maintained.

About 35 years ago, Mather (1943) stated that the function of polygenes might be regulatory and important for adaptation and speciation. Recently, Mukai and Cockerham (1977) obtained experimental results suggesting that most of the viability polygenes are located outside the structural genes. Their argument is based on the high mutation rates of viability polygenes on a chromosome basis in comparison with the total mutation rate of structural genes which was estimated using the mutation rates of isozyme genes in *D. melanogaster*. Thus, it is a reasonable speculation that most of the viability polygenes are some kind of regulatory genes, and that they play an important role in the evolution of organisms.

REFERENCES

Bateman, A.J. (1959). *Int. J. Radiation Biol. 1*, 170-180.
Burdick, A.B., and Mukai, T. (1958). *Proc. 2nd Int. Congr. Peaceful Uses of Atomic Energy, Geneva*, 325-329.
Bonnier, G., and Jonsson, U.B. (1957). *Hereditas 43*, 441-461.
Clayton, G., and Robertson, A. (1955). *Amer. Natur. 89*, 151-158.

Clayton, G.A., and Robertson, A. (1964). *Genet. Res., Camb.* 5, 410-422.
Comstock, R.E., and Robinson, H.F. (1952). In "Heterosis" (J. Gowen, ed.), pp. 494-516. Iowa State College Press, Ames.
Darlington, C.D., and Mather, K. (1949). "The Elements of Genetics." George Allen and Unwin Ltd., London.
Durrant, A., and Mather, K. (1954). *Genetica* 27, 97-119.
Fisher, R.A., and Yates, F. (1938). "Statistical Tables." Oliver and Boyd, Edinburgh.
Friedman, L.D. (1964). *Genetics* 49, 689-699.
Greenberg, R., and Crow, J.F. (1960). *Genetics* 45, 1153-1168.
Haldane, J.B.S. (1956). *J. Genet.* 54, 294-296.
Käfer, E. (1952). *Z. Ind. Abst. Vererb.* 34, 508-535.
Kerkis, J. (1935). *Summ. Commun. 15th Int. Physiol. Congr.*, 198-200.
Kerkis, J. (1938). *Bull. Acad. Sci. URSS*, 75-96 (Russian with English summary).
Kimura, M. (1965). *Proc. Natl. Acad. Sci., U.S.A.*, 54, 731-736.
Kitagawa, O. (1967). *Japan J. Genet.* 42, 121-137.
Lüning, K.G., and Jonsson, S. (1957). *Proc. 5th Int. Conf. Radiobiology, Stockholm 1956*, 2, 425-432.
Maruyama, T., and Crow, J.F. (1975). *Mut. Res.* 27, 241-248.
Mather, K. (1943). *Biol. Rev.* 18, 32-64.
Mather, K., and Wigan, L.G. (1942). *Proc. Roy. Soc., B, 131*, 50-64.
Mukai, T. (1964). *Genetics* 50, 1-19.
Mukai, T. (1970). *Genetics* 65, 335-348.
Mukai, T. (1977). In "Proceedings of the Second Taniguchi International Symposium on Biophysics: Molecular Evolution and Polymorphism" (M. Kimura, ed.), pp. 103-126.
Mukai, T., and Cockerham, C.C. (1977). *Proc. Natl. Acad. Sci., U.S.A.*, 74, 2514-2517.
Mukai, T., and Yamaguchi, O. (1974). *Genetics* 76, 339-366.
Mukai, T., and Yamazaki, T. (1968). *Genetics* 59, 513-535.
Mukai, T., Chigusa, S., and Yoshikawa, I. (1964). *Genetics* 50, 711-715.
Mukai, T., Yoshikawa, I., and Sano, K. (1966). *Genetics* 53, 513-527.
Mukai, T., Chigusa, S.I., Mettler, L.E., and Crow, J.F. (1972). *Genetics* 72, 335-355.
Mukai, T., Cardellino, R., Watanabe, T.K., and Crow, J.F. (1974). *Genetics* 78, 1195-1208.
Muller, H.J. (1934). *Verhandl. 4 Intern. Kongr. Radiol. (Zürich)* 2, 100-102.
Muller, H.J. (1950). *Amer. J. Human Genet.* 2, 111-176.

Ohnishi, O. (1977). *Genetics 87*, 529-545.
Paxman, G.J. (1957). *Genetica 29*, 39-57.
Robertson, A. (1956). *J. Genet. 54*, 236-248.
Rokizky, P. (1936). *Usp. Dootekhn. Nauk. 2*, 161-202 (from Clayton and Robertson, 1955).
Scossiroli, R.E. (1954). *Atti IX cong. Int. Genet. Caryologia 4 (Suppl.)*, 861-864.
Scossiroli, R.E., and Scossiroli, S. (1959). *Int. J. Radiation Biol. 1*, 61-69.
Serebrovsky, R.E. (1935). *Zool. Zh. 14*, 465-480 (from Clayton and Robertson, 1955).
Simmons, M. (1976). *Genetics 84*, 353-374.
Timoféeff-Ressovsky, N.W. (1935). *Mathematisch-physikalishe Klasse. Fachgruppe VI. Biologie. Neue Folge 1*, 163-185.
Tobari, I., and Nei, M. (1964). *Genetics 52*, 1007-1015.
Wallace, B. (1956). *J. Genet. 54*, 280-293.
Wallace, B. (1957). *Proc. Natl. Acad. Sci., U.S.A., 43*, 404-407.
Wallace, B. (1958). *Evolution 12*, 532-556.
Wallace, B. (1959). *Proc. 10th Int. Congr. Genet. 1*, 408-419.
Watanabe, T.K., Yamaguchi, O., and Mukai, T. (1976). *Genetics 82*, 63-82.
Yamada, Y. (1961). *Japan J. Genet. 36 (Suppl.)*, 78-87.
Yamada, Y., and Kitagawa, O. (1961). *Japan J. Genet. 36*, 76-83.

USES OF RECOMBINANT INBRED LINES

Alberto Oliverio

Istituto di Fisiologia Generale
University of Rome
and
Laboratorio Psicobiologia e Psicofarmacologia
Rome, Italy

I. AN HISTORICAL OVERVIEW OF RI STRAINS IN RELATION TO TISSUE TRANSPLANTATION

More than ten years ago Roderick and Schlager (1966) published a review on multiple factor inheritance dealing with the methods of analyzing quantitative characteristics and the mechanisms by which allelic differences have small individual effects on phenotypic variation. As it was pointed out, genes may have large effects and be called "major genes" or they may have individually small effects and be called "minor genes". The distinction between major and minor genes was indicated to be rather arbitrary, since the effect of individual allelic differences on the phenotypic variation may range from small to very large. It is useful, however, to assess whether the genotypic variance of a trait is caused by allelic differences at one or two loci or by allelic differences at several loci. As an example we might consider body weight of mice in which phenotypic variation is due to segregation of alleles with minor effects at many loci, though major genes such as the *obese* (*ob*) mutant may exert a clear-cut effect.

Different genetic methods have been devised in order to give measures of quantitative characters, to give estimates of the number of loci governing a given trait, to identify specific loci, and to locate new genes by conventional linkage tests. If we look back to classic textbooks, such as Mather (1949), Reeve and Waddington (1952), Falconer (1960), or to some chapters of Green (1966), and if we go through the different sections of these volumes, we may realize how many impressive experiments have been conducted in recent years and we can see that much information has been collected using different genetic strategies. Quantitative genetics using the laboratory mouse has reached a relevant place, as indicated by the number of findings reported in *Mouse News Letter*.

An analytic genetic device called strain distribution pattern (SDP) has been developed in recent years by Bailey (1965) and by Bailey and Mobraaten (1969) by using a set of recombinant inbred (RI) strains, though similar use of SDPs under genetically less controlled situations was made earlier for identifying blood group loci (Amos *et al.*, 1963) and for defining antigenic specificities at the H-2 locus (Snell and Stimpfling, 1966). The use of RI strains simultaneously exploits the advantages offered by genetic segregation and recombination on one hand and of inbreeding on the other hand. Genetic segregation and recombination have revealed the particulate nature of genes and the characteristics of linkage, while inbreeding, which represents the opposite of segregation and recombination, represents the means of producing replicable genotypes. To smooth the path of the reader we shall first define RI strains before entering into the details of their history and of their uses.

RI strains, as they were defined by Bailey (1971), are those which have been derived from the cross of two unrelated but highly inbred progenitor strains and which have been maintained independently under a regimen of strict inbreeding since the F_2 generation. This procedure genetically fixes the chance recombinants that occur in the generations following the F_1 in ever decreasing amounts as full homozygosity is approached. The resulting population of strains can be looked upon as a replicate recombinant population.

The research on SDP and RI strains derives from studies on the genetics of tissue transplantation and of histocompatibility; however, as we shall see later, RI strains were later used in a number of other genetic analyses. Since we shall deal with problems of immunogenetics and tissue transplantation we shall define some key terms which may not be familiar to some readers.

1. *Histocompatibility* refers to the growth or failure to grow of tissue transplants. Thus, histocompatibility genes are the genes that determine susceptibility and resistance to transplants. In general terms the histocompatibility loci are those whose normal function is to produce the chemical building blocks from which the cell membrane is formed. Their role is obviously relevant not only to basal conditions but also

in a number of immunological processes and in tumorigenesis.

2. *Isogenic* indicates genetic identity of animals or tissues. The members of an inbred strain are regarded as isogenic.

3. *Coisogenic* strains are strains genetically identical except for a difference at a single locus. True coisogenicity may not be achieved and those strains which are approximately coisogenic are called *congenic*.

4. An *isogenic graft* is a graft between genetically identical individuals, namely between animals of a given inbred strain or between the F_1 hybrids produced by crossing inbred strains, while an *allogenic graft* is a graft between genetically disparate individuals.

5. An *alloantigen*, as defined by Snell and Stimpfling (1966), is the product of a heteromorphic locus (a locus existing in two or more allelic forms) such that alternative forms are antigenic in individuals lacking them. It incites an immune response when transferred within species.

Following this short terminological digression we shall review the history of the application of recombinant inbreeding to specific genetic analyses. As previously noted Snell and Stimpfling (1966) described the pattern of distribution of histocompatibility alleles and alloantigenic specificities among inbred strains and congenic resistant sublines (Table I). Comparable information was given for the H-2 system. The tendency of lines known to be related to have similar histocompatibility genotypes (*e.g.*, C57BL/10, C57Br/cd, and C58; A and BALB/c; DBA/1 and DBA/2) was first noted. Second, evidence concerning the multiplicity of alleles was considered and it was suggested that a minimum of fifteen loci govern the compatibility of tissue grafts.

The use of SDPs under genetically more controlled situations was made a few years later by Bailey (1965; Bailey and Mobraaten, 1965). In a study dealing with the search for influences of genetic background on survival time of skin grafts from mice bearing Y-linked histoincompatibility, a sample of four from nine inbred lines independently derived from matings *inter se* of the F_1 hybrids (BALB/c females × C57BL/6 males) designated as CBF_1 were used. These lines, designated CXBD, CXBE, CXBH and CXBK, were genetically differentiated owing to chance fixation of genes during their subsequent independent maintenance under full-sib mating (Note: the proper symbol for RI strains employs a capital letter X, to avoid confusion of RI lines with an F_1 hybrid). Males of these lines carried Y chromosomes derived solely from their C57BL/6 male ancestor. Following the genetic laws of transplantation, skin grafts from such males would be compatible when placed on the CBF_1

TABLE 1. Distribution of histocompatibility alleles and alloantigenic specificities among inbred strains and congenic resistant sublines (from Snell and Stimpfling, 1966).

Strain	Allele					Alloantigenic specificities				
	H-1	H-3	H-7	H-5	H-6	Kappa	Iota	TL	θ-AKR	θ-C3H
A	not-c	not-a	not-a	+	+	–	+	+	–	+
AKR	not-c	not-a	a					–	+	–
BALB/c	not-c	not-a	not-a	–	+	–	+	–	–	+
C3H/He	a	not-a	not-a					–	–	+
C3H.K	b	not-a	not-a							
C3H/St				+						
C57BL	c	a	a	–	+	–	+	–	–	+
B10.BY	b?	a	a							
B10.C(47N)	c	a	not-a							
B10.129(5M)	b	a	a							
B10.129(13M)	c	not-a	a							
B10.129(21M)	c	a	a	–	+					
C57BR/cd	c	a	a							
C57I	c	a	a							
C58	c	not-a	a	–	+	–	–	+	–	+
DBA/1		not-a	a	–	–	–	+		–	+
DBA/2	d	not-a	a	–	–	–	+		–	+
F/ST						–	–			
RF						+		+	+	–
SWR						–	+		–	+
YBR/HeHa					+	+	+			
129	b?	not-a	a	+	+	–	+		–	+

female host in regard to effects of all dominant H genes except those of the Y chromosome. Thus, every male should have borne the same source of histoincompatibility toward the female host, but males of different lines would concomitantly carry different genetic backgrounds. The study was based on tests of the survival on CBF_1 females of male skin graft of two inbred lines (Table II). Though the differences between the lines were not significant, suggesting that the genetic background produced no apparent influence on the efficacy of the Y antigen to elicit graft rejection (for only one introduced source of histoincompatibility) this represents the first attempt to find an SDP.

The above study was based upon four RI lines only and these strains reached generation F_9-F_{11} of inbreeding; however, a larger number of RI lines were available and a few years later the progress of inbreeding in nine different RI lines was reported (Bailey and Mobraaten, 1969). In fact, an attempt at estimating the number of histocompatibility loci was based on survival of grafts exchanged between sibs of nine partially inbred (F_{11}) RI lines and approximate minimal estimates of the number of loci were calculated within each RI strain.

When the RI lines reached approximately the twentieth generation of inbreeding, their characteristics were reported. It was of interest that only one out of seven strains tested

TABLE II. *Means and standard deviations of the time elapsed (days) since grafting for the onset of rejection of male grafts on CBF_1 female hosts (modified from Bailey, 1965).*

Line	Generation	No. Hosts	Mean[a]	S.D.	t[b]
CXBE	F_9	7	16.86	1.57	0.29
CXBH	F_9	9	16.44	3.97	
CXBD[c]	$F_9+F_{10}+F_{11}$	22	14.82	2.34	1.07
CXBK[c]	F_9+F_{10}	25	15.86	3.04	

[a]*Unity has been subtracted from the day of rejection as correction for discontinuity.*

[b]*None of these values of t are significant at the 10% level.*

[c]*Data pooled from three different series of tests.*

TABLE III. Designations and characteristics of eight RI lines derived from a cross of BALB/cAn females and C57BL/6JN males (C and B strains, respectively) (from Bailey and Mobraaten, 1969).

Strain	Generation inbreeding	Genetics of coat pigment	Histocom. fixed	Female skin compat. on CBF_1 female	Strain origin of X-linked compat.
CXBD	F_{22}	++++++	No	No^a	C (probably)
CXBE	F_{21}	aa++++	No	Yes	B
CXBF	F_{19}	++bbcc	No	Yes/No^b	C
CXBG	F_{23}	++bbcc	No	Yes/No^b	C
CXBH	F_{23}	aabb++	Yes	Yes	B
CXBI	F_{20}	(B?)cc	No	Yes	C
CXBJ	F_{21}	aabb++	?	Yes	?
CXBK	F_{20}	aa++++	No	Yes	C (probably)

^{a}H mutation arisen and fixed since F_{10}.

^{b}Weak H mutation still segregating.

had their histocompatibility (H) genes fixed by that stage of inbreeding and that mutations had occurred. Table III summarizes these data. It is interesting to note that the situation has evolved and that at December 31, 1977, after more than 60 generations of full-sib mating, there were 7 RI lines (CXBF is extinct), the H genes were fixed in all lines, and no mutations were present in the lines with the exception of line CXBD.

II. POSSIBLE APPLICATIONS OF RI STRAINS TO QUANTITATIVE GENETIC ANALYSES

As we have already underlined the interest in RI lines derived from studies on the genetics of histocompatibility and the group of the previously described RI strains served as an aid to identifying individual lines in a battery of newly developed B6 congenic H gene lines. As noted by Bailey (1971) the congenic lines must not be confused with ther RI strains themselves. It will suffice to say that each of the congenic lines was independently developed from an initial cross of B6 by a regimen of skin graft testing and backcrossing to B6 for at least 12 generations. This procedure resulted in a battery of congenic B6 lines, each differing from B6 itself by only an introduced C strain allele chromosomal segment carrying a distinctive allele at a single H locus.

Each of the congenic H gene lines were tested against the RI strains to find which of the RI strains carry the C strain allele and which the B6 allele, at that particular H locus. This was done by grafting skin from donors of the congenic lines onto the F_1 hosts which were produced by crossing each of the RI strains with the B6 strain. A surviving graft indicated that the RI strain, parental to that F_1 host, carried the C strain allele while a rejected graft indicated the B6 strain allele. The resulting strain distribution patterns for a sample of eight different H loci are presented in Table IV.

What are the experimental applications of the RI strains? At the time of the study previously reported (Bailey, 1971), four distinct possibilities were envisaged.

(1). The first application was related to the identification of H loci. There is the possibility that new congenic H-gene lines will duplicate one another or existing congenic H lines. Although this situation can be tested directly by exchanging skin grafts between congenic lines in all combinations, it was suggested that this may be done more quickly by finding the SDP of the newly isolated gene in the RI strains. Identical patterns suggest that the locus is already represented (in order to exclude matching by chance, a graft exchange between the new line and its SDP-matching line is required). On the contrary different SDPs can be taken as good evidence that two different loci are involved.

(2). A second application was related to a search for H gene functions. It was suggested that the RI strains might also be useful for finding pleiotropic effects of the H genes. It was, in fact, hypothesized that the immune response genes are associated with major H loci. In this line the localization of a gene controlling the female immune response to the male antigen has been reported (Bailey and Hoste, 1971).

(3). A third application was related to the analysis of traits that depend on replicative observations and require measurements on several mice of the same genotype. Since the evaluation of such traits cannot be done in genetically segregating populations usually employed in genetic analyses, the use of RI strains might aid in the evaluation of the controlling gene system and suggest possible crosses in order to

TABLE IV. RI SDPs of BALB/c (C) and C57BL/6 (B6) alleles at three coat color loci and eight histocompatibility loci (from Bailey, 1971).

locus[a]	SDPs in RI strain						
	CXBD	CXBE	CXBG	CXBH	CXBI	CXBJ	CXBK
a	C	C	C	B6	B6	B6	B6
b	B6	B6	C	C	C	C	B6
c	B6	B6	C	B6	C	B6	B6
H(w13)	C	B6	C	C	C	B6	B6
H(w17)	C	C	C	C	C	B6	C
H(w19)[H-2][b]	C	B6	B6	C	B6	B6	B6
H(w20)	B6	C	B6	C	B6	C	B6
H(w35)	B6	C	C	C	C	C	B6
H(w38)	B6	C	B6	C	C	C	B6
H(w80)[H-1][b]	B6	B6	C	B6	C	B6	B6
H(w96)	C	C	B6	B6	C	C	C

[a]Tentative designations of H loci are indicated by the letter w and a number enclosed in parentheses to avoid confusion with names of established H loci.

[b]Recently identified by F tests as the H locus indicated in brackets. The congenic lines distinguishing H(w19) and H(w80) will hereafter be designated as B6.C-H-2^d and B6.C-H-1^b, respectively

isolate the effective genes.

(4). Finally, a fourth application was envisaged, the selection of marker genes to succeed in locating a new gene by conventional linkage tests. When the H genes have been located on the linkage map (but other genes are, of course, also suitable), they can become markers. It was indicated that the RI strains might become useful in selecting the H gene markers most likely to be successful in locating a new gene. The new genes with which the method can be used are only those by which the B6 and C strains differ, namely the extant polymorphic genes, such as those determining isozymes, immunoglobulins, and alloantigens.

III. APPLICATION OF RI STRAINS TO THE ANALYSIS OF DIFFERENT PHENOTYPES

In 1971 the RI strains were typed as carrying either the C or B6 allele at each of 22 H-loci and at each of more than 10 loci that determine isozymes, hemoglobin, immunoglobin, alloantigens, and coat color. At the end of 1977 SDPs of 70 marker genes were available, the resulting SDPs being useful in searching for gene identity, since identical or similar SDPs of any pair of traits indicate possible pleiotropism or linkage of controlling genes. The utility of this method increases as the SDP list expands. In addition, since the number of RI lines derived from crosses of inbred strains other than C and B6 has also increased (Taylor et al., 1971), it may be possible to determine not only gene segregation but also gene linkages by typing each RI line for a new marker and comparing the SDP with those already available for other genes.

Since 1970 inbreeding of a number of recombinant lines derived from F_2 generations of crosses between strains AKR and C57BL (16 RI lines), C57BL/6J × DBA/2J (25 RI lines), C57BL/6J × SJL/J (2 RI lines), and C57BL/6J × AKR/J (1 RI line) was carried out (Taylor et al., 1971, 1973a, 1976, 1977). The availability of these RI lines offers larger research possibilities than a single set of RI lines derived from two strains only in that several miscellaneous lines derived from different crosses may be informative with respect to linkage of given loci. An example of this opportunity is given by the use of multiple RI lines derived from crosses between strains permissive to N- and B-tropic murine leukemia viruses (Fv-1) in a quantification of the linkage relationships between Fv-1 and hexose-6-phosphate dehydrogenase isozyme locus (Gpd-1) (Taylor et al., 1977). By using this procedure it was shown that the Fv-1 and Gpd-1 loci were inherited concordantly in all but a single case, confirming the close linkage between those two loci. None of several miscellaneous RI lines informative with respect to the Gpd-1 and Fv-1 linkage in fact involved recombination between Fv-1 and Gpd-1 (Table V).

As we have seen the applications of the SDP are extremely versatile and increase as the SDP list expands. In a search for the distribution and linkage of the locus controlling structure of the y-chain of mouse embryonic hemoglobins (Hbb) 115 inbred stocks of mice were selected, 41 being standard inbred strains, 24 being mutant-bearing stocks, 4 being multiply-marked Linkage Testing Stocks and 46 being different RI lines. These latter lines provided an opportunity for the calculation of the upper limits of recombination frequency of Hbb (Stern et al., 1976).

TABLE V. Fv-1 and Gpd-1 types of some miscellaneous RI lines (from Taylor et al., 1977).

RI line[a]	Genotype	
	Gpd-1	Fv-1
BxJ-1	b	n
BxJ-2	b	n
HP/Ei	a	b
LT/Re	b	b
SEA/Gn	a	n
TSK/Le	a	b

[a] The BxJ-1 and BxJ-2 RI lines were derived from crossing C57BL/6L($Gpd-1^a$ $Fv-1^b$) with SJL/J ($Gpd-1^b$ $Fv-1^n$). HP/Ei is an RI strain derived from C57BL/6J and AKR/J ($Gpd-1^b$ $Fv-1^n$). Lt/Re and SEA/GnJ are RI strains derived from crossing BALB/c ($Fv-1^b$ $Gpd-1^b$) with C58($Fv-1^n$ $Gpd-1^a$) and P/J ($Fv-1^n$ $Gpd-1^a$), respectively. TSK/Le is an RI strain derived from crossing B.10.D2(58N)/Sn ($Fv-1^b$ $Gpd-1^a$) with C3H/Di ($Fv-1^n$ $Gpd-1^b$).

From their initial main application in the field of tissue transplantation and histocompatibility, the use of RI strains has now spread to many distinct fields of biology. At the end of 1977 the publications dealing with RI lines amounted to a total of 87 papers. It is perhaps useful to divide them by field of interest and to refer to all the studies in the field. A tentative subdivision of these fields, though many overlappings are possible, indicates seven main categories.

A. *Immunogenetics*

A large number of such experiments have been devoted to genetic analyses of histocompatibility or to mapping genes within the H complex (Arnaiz-Villena et al., 1976; Bailey, 1965, 1971; Chesebro et al., 1974; Dorf et al., 1975; McKenzie, 1975; Merryman and Maurer, 1975; Merryman et al., 1972, 1975; Mobraaten et al., 1973; Shultz and Bailey, 1975). A number of immune responses to bacteria, viruses or specific bacterial biochemicals, and of markers regulating natural antibody production have also been studied (Bailey and Hoste,

1971; Baker et al., 1976a, 1976b; Bishop et al., 1976; Brown and Crandall, 1976; Cramer and Riviere, 1960; De Maeyer et al., 1975; Eichmann and Nisonoff, 1975; Imanishi and Makela, 1974; Lieberman et al., 1974a, 1974b, 1975, 1976a; Martelly et al., 1972; McCarthy et al., 1976; Morse et al., 1977; Mushinski and Potter, 1975; Potter et al., 1973; Riblet et al., 1975; Snell et al., 1973; Taylor et al., 1971, 1973b, 1975a, 1975b; Watson et al., 1977).

B. *Tumors and Related Immune Reactions*

Analyses of chemically-initiated tumorigenesis of immunological reactions connected to myeloma proteins or of susceptibility to plasmocitoma have been conducted (Aoki et al., 1973; Atlas et al., 1976; Berek et al., 1976; Herbeman and Aoki, 1972; Lieberman et al., 1975, 1976b; Morse et al., 1976; Potter et al., 1975; Thomas et al., 1973).

C. *Biochemicals and Enzymes*

Plasmatic serotonin on corticosterone levels, galactokinase activity, embryonic hemoglobins or adrenal lipid depletion have been analyzed through the use of RI strains (Eleftheriou and Bailey, 1972a, 1972b; Gurtoo et al., 1977; Haugen et al., 1976; Meisler, 1976; Mishkin et al., 1976; Russel et al., 1972; Stern et al., 1976; Taylor and Meier, 1976; Taylor et al., 1974).

D. *Central Nervous System*

Hypothalamic norepinephrine, regional brain noradrenaline, turnover levels or corticosterone retention, cortex weight or hormonal effects on brain function have been assessed (Eleftheriou, 1974a, 1974b; Eleftheriou et al., 1975, 1976; Eleftheriou and Kristal, 1974; Kempf et al., 1978; Moisset, 1977a, 1977b).

E. *Behavior*

A number of behavioral traits ranging from aggression to learning have been studied (Eleftheriou et al., 1974; Lucas and Eleftheriou, 1975; Messeri et al., 1975; Oliverio et al., 1973a, 1973b; Simmel et al., 1977; Taylor, 1976a, 1976b). (See the next section for further discussion of this subject.)

F. *Psychopharmacology*

Genetic differences in opiate receptors and genetic analyses of the effect of opiates, alcohol, anticholinergic or adrenergic drugs have been reported (Baran *et al.*, 1975; Castellano *et al.*, 1974; Elias and Eleftheriou, 1975; Oliverio *et al.*, 1973b, 1975; Shuster *et al.*, 1975a, 1975b, 1977).

G. *Miscellaneous Fields*

Genetic analyses based on the use of RI strains were also applied to problems such as cadmium toxicology (Taylor *et al.*, 1973a) and skeletal mutations (Varnum and Stevens, 1974).

IV. BEHAVIOR GENETIC ANALYSES

I will take the opportunity presented by behavior genetics in order to discuss in more detail the uses of RI lines to investigate the effects of small numbers of genes and the limitations of this technique, in the context of other quantitative genetic methods.

As previously noted RI lines derived from C and B6 strains have been employed in behavior genetic or psychopharmacogenetic analyses of behavior patterns such as exploratory behavior, dominance and fighting, avoidance and maze learning or the effects of adrenergic, anticholinergic agents, alcohol or opiates upon behavior. We may subdivide these different studies on the ground of the hypothesis regarding possible genetic control of different behavioral patterns in mice.

A first category is represented by investigations such as those on dominance behavior (Messeri *et al.*, 1975), maze learning (Oliverio *et al.*, 1975b) or morphine sensitivity (Oliverio *et al.*, 1975a; Shuster *et al.*, 1975b), in which the groupings of the various RI strains suggested a polygenic genetic model based on at least two, three, and possibly many more loci. For example, if we consider dominance behavior, the analysis of the properties of the statistical groupings of the progenitor and RI strains and the F_1 hybrids, suggested that the simplest genetic hypothesis is one which assumes that there are at least three loci responsible for this behavior. This assumption was based on the statistical groupings of the various strains. Each distinctly different phenotype must have a distinctive genotype, assuming a common environment. The number of loci required to explain the number of genotypes depends on the number of combinations of alleles and loci that

one can find. In the case of dominance behavior, due to the
manner of derivation of the RI strains, these must be homozygous. Thus, five distinct phenotypes were apparent, so that
the assumption was made that this behavior was influenced by
at least three loci.

A second category is represented by studies such as those
on active avoidance behavior, on the effects of alcohol on
motor activity or on exploratory activity, and on the effects
of scopolamine -- a cholinergic drug -- upon the behavior.
In these studies the genetic model is "simpler" and it was
possible to characterize a gene exerting a major effect on
that specific behavior. If we take as an example exploratory
behavior, analyses of the data obtained in the testing of all
strains indicated that these data were distinctly grouped into
two major categories: strains CXBD, CXBI and CXBJ, and the
progenitor strain BALB/c exhibited low basal exploratory activity while all other RI strains, the two reciprocal F_1 hybrids
and the other progenitor strain, C57BL/6, were high in basal
exploratory activity (Table VI). The RI lines' SDP closely
matched that for an existing line, H(w26), which was found to
exhibit low exploratory activity, thus confirming the single
major gene hypothesis for exploratory activity.

It is worth emphasizing the strength of evidence provided
when it is shown that a B6 congenic line differs from B6, the
recurrent parental strain. No matter if the proof is unconventional, it is nevertheless more stringent proof of linkage
than is the data obtained in most of the more conventional
linkage tests. This is so, because of the backcrossing derivation of the congenic lines. The matching or near matching
of SDPs results, with luck, in finding two or three candidate
congenic lines, though more often only one or none are found.
Thus, biased selection of data is not possible. Even if we
looked at all 25 of our congenic lines, however, the following
argument shows that the evidence of linkage would still be
strong. The probability that one of the lines carries a particular BALB/c allele (unlinked to the differential H locus)
is $(\frac{1}{2})^g$, where g is the number of backcross generations (beyond F_1) used in establishing the congenic line. The probability (P) that at least one of the N tested candidate lines
carries (unlinked) the particular allele of interest is:

$$1 - \left[1 - (\tfrac{1}{2})^g\right]^N.$$

If g is 11 (see Bailey, 1971, page 325) and N is 3, P = 0.0015,
and if N is even as high as 25, P is still only 0.012. Thus,
although unconventional, the evidence that the gene of interest is linked to the differential H gene is strong.

TABLE V. Exploratory activity (mean number of shuttle box crossings + S.E.M.) in BALB/c and C57BL/6, their reciprocal F_1 hybrids, $B6CF_1$ and $CB6F_1$, seven recombinant inbred strains (CXBD, CXBE, CXBG, CXBH, CXBI, CXBJ, CXBK) and congenic line H(w26) necessary to confirm the genetic model. All strains are arranged in statistically significant rank order of low to high response (from Oliverio et al., 1973a).

Rank order of strains[a]	Strains	Exploratory Activity
Low	CXBI	8.8±2.1
	CXBJ	8.8±1.1
	BALB/c	13.4±0.9
	CXBD	16.7±0.7
	H(w26)	11.8±1.14
	$B6CF_1$×BALB/c	13.4±0.62
High	CXBG	22.4±2.2
	CXBE	22.7±1.0
	$CB6F_1$	23.2±1.2
	CXBK	23.5±1.8
	CXBH	23.7±1.4
	$B6CF_1$	24.8±1.2
	C57BL/6	27.1±1.1
	$B6CF_1$×BALB/c	28.4±0.8

[a] Based on one-way analysis of variance, Student-Newman-Keuls and Tukey's W, low and high groups are significantly different at the 0.01 level.

For exploratory activity, evidence for a major gene hypothesis was also based on backcross data. The backcross $B6CF_1$ ×BALB/c mice were seen to comprise two distinct and statistically separate groups (Figure 1), indicating that exploratory activity is controlled by a single gene exerting a major effect on this behavior. The symbol *Exa* was suggested to designate the locus, linked to H(w26) in chromosome 4 (LGVIII) for basal short-term exploratory activity, with Exa_l^h to designate the allele determining high activity, and Exa^l the

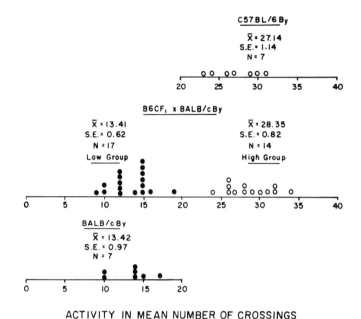

FIGURE 1. *Distribution and mean activity in backcross B6CF₁×BALB/c. On the ground of statistical calculations of the probability that each backcross observation resembled one parental strain or the other, 19 was considered to be an observation belonging to the lower group and 24 was considered an observation belonging to the high group. (from Oliverio et al., 1973a).*

allele determining low level of activity.

A similar analysis, conducted for active avoidance learning, allowed characterization of a single dominant autosomal gene which influences this behavior (Figure 2). Use of congenic lines B6.C-H-25^c and B6.C-H(w56) determined linkage of this gene *Aal* on chromosome 9 (LGII) of the mouse. In general these findings show that it is possible to ascribe to a single gene a major effect on a complex behavioral trait such as avoidance learning or on other behavioral patterns. The isolation of the genes influencing these behavioral units in extant congenic lines allows for more definitive studies on many behavioral phenomena and their physiological and biochemical correlates.

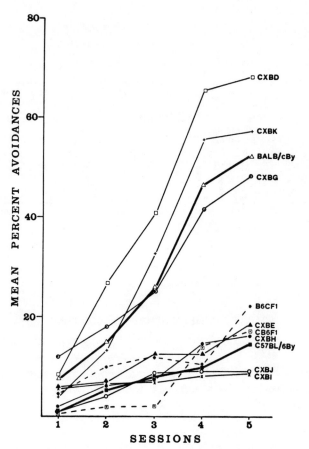

FIGURE 2. Mean percent avoidances per session (Day 1-5) for seven RI lines, their two progenitor strains, and two reciprocal F_1 hybrids. (from Oliverio et al., 1973b).

V. USES AND LIMITATIONS OF RI STRAINS

What is the meaning of the findings on behavior and, in general terms, what is the role played by "major" genes on complex traits? If we consider again the example offered by a genetic analysis of exploratory activity or by active avoidance learning we see that the analyses based on the use of RI strains seem to conflict with two previous findings.

First, for active avoidance learning, a previous study based on Matherian analyses of dominance and segregation led to an estimate that active avoidance behavior was controlled by not less than three (C57BL/6×DBA/2 hybrids) and probably not less than four (C57BL/6×SEC/1Re hybrids) segregating units (Oliverio et al., 1972).

A second genetic approach was based on the use of 31 different single-gene mouse mutants on the inbred C57BL/6 background tested for exploratory activity and active avoidance learning (Oliverio and Messeri, 1973). The effects of 27 coat-color alleles and four other mutations were assessed by comparing the performance of mutant mice to that of their siblings (heterozygous or normal) and to that of a group of normal C57BL/6 mice. Twelve mutants exhibited exploratory levels higher or lower than that of normal mice and seven mutants were found to attain an avoidance performance level significantly higher than that evident in control mice.

Despite these findings the results obtained by using the RI approach support the thesis that a single major gene influences exploratory activity (or active avoidance) performance in mice, even though other genes that may contribute to the genetic variance of this trait may be present. This is a point which must be kept in mind since characterization of a single major gene and its linkage do not indicate that that gene is the sole one responsible for genetic variation. A major gene may in fact contribute substantially to the genetic variance of a given trait, particularly of a behavioral phenotype, thus concealing the effects of other minor genes. It is also important to remember that a number of environmental factors contribute to the variation of a behavioral phenotype, by acting on behavioral development or on the behavior of adult animals (Oliverio, 1977).

However, the advantages of isolating a given major gene in extant congenic lines are related to the possibility of assessing the role of its biochemical correlates on essential motivational systems or, at least, to form a "catalog" of the different biological factors which modulate a given behavior. A more optimistic and clear-cut approach is related to the use of RI strains in relation to the study of psychotropic drugs or to the analysis of biochemical variants: in this instance it is possible to identify and map the gene(s) responsible for drug action or for given biochemical variants and to clarify specific biochemical steps. Examples of this approach are given by different findings on mammalian biochemical genetics, such as those on androgen inducibility of kidney β-glucouronidase (Swank and Bailey, 1973), on intestinal aminopeptidase (Womack et al., 1975) or embryonic hemoglobin variants (Stern et al., 1976). These examples indicate that an important genetic tool is available for assessing a number of biochemical and biological correlates of gene action and that RI strains represent a useful device in the study of quantitative genetic variation.

ACKNOWLEDGMENTS

I wish to thank Dr. D.W. Bailey for his advice and criticism. Dr. Joan Staats, the Jackson Laboratory, Bar Harbor, has helped me with the search of the literature.

REFERENCES

Amos, D.B., Zumpft, M., and Armstrong, P. (1963). *Transplantation* 1, 270.
Aoki, T., Potter, M., and Sturm, M.M. (1973). *J. Nat. Cancer Inst.* 51, 1609-1617.
Arnaiz-Villena, Taylor, B.A., and Festenstein, H. (1976). *Folia Biol. (Prague)* 22, 379-380.
Atlas, S.A., Taylor, B.A., Diwan, B.A., and Nebert, D.W. (1976). *Genetics 83*, 537-550.
Bailey, D.W. (1965). *Transplantation 3*, 531-534.
Bailey, D.W. (1971). *Transplantation 11*, 325-327.
Bailey, D.W., and Hoste, J. (1971). *Transplantation 11*, 404-407.
Bailey, D.W., and Mobraaten, L.E. (1965). *Mouse News Letter 32*, 76-77.
Bailey, D.W., and Mobraaten, L.E. (1969). *Transplantation 7*, 394.
Baker, P.J., Amsbaugh, D.F., Prescott, B., and Stashak, P.W. (1976a). *J. Immunogenet.* 3, 275-286.
Baker, P.J., Amsbaugh, D.F., Prescott, B., and Stashak, P.W. (1976b). *Folia Biol. (Prague)* 22, 407-408.
Baran, A., Shuster, L., Eleftheriou, B.E., and Bailey, D.W. (1975). *Life Sci.* 17, 633-640.
Berek, C., Taylor, B.A., and Eichmann, K. (1976). *J. Exp. Med.* 144, 1164-1174.
Bishop, C., Festenstein, H., and Taylor, B.A. (1976). *Folia Biol. (Prague)* 22, 422.
Brown, A.R., and Crandall, C.A. (1976). *J. Immunol.* 116, 1105-1109.
Castellano, C., Eleftheriou, B.E., Bailey, D.W., and Oliverio, A. (1974). *Psychopharmachologia 34*, 309-316.
Chesbro, B., Wehrly, K., and Stimpfling, J.H. (1974). *J. Exp. Med. 140*, 1457-1467.
Cramer, R., and Riviere, M.R. (1960). *Bull. du Cancer 47*, 89-95.
DeMaeyer, E., DeMaeyer-Guignard, J., and Bailey, D.W. (1975). *Immunogenet.* 1, 438-443.

Dorf, M.E., Plate, J.M.D., Stimpfling, J.H., and Benacerraf, B. (1975). *J. Immunol. 114*, 602-605.
Eichmann, K., and Nisonoff, A. (1975). *Immunogenet. 1*, 526-527.
Eleftheriou, B.E. (1974a). *Brain Res. 70*, 538-540.
Eleftheriou, B.E. (1974b). *Brain Res. 69*, 77-82.
Eleftheriou, B.E., and Bailey, D.W. (1972a). *J. Endocrinol. 55*, 225-226.
Eleftheriou, B.E., and Bailey, D.W. (1972b). *J. Endocrinol. 55*, 414-420.
Eleftheriou, B.E., and Kristal, M.B. (1974). *J. Reprod. Fertil. 38*, 41-47.
Eleftheriou, B.E., Bailey, D.W., and Denenberg, V.H. (1974). *Physiol. Behav. 13*, 773-777.
Eleftheriou, B.E., Elias, M.F., Castellano, C., and Oliverio, A. (1975). *J. Hered. 66*, 207-212.
Eleftheriou, B.E., Elias, M.F., Cherry, C., and Lucas, L.A. (1976). *Physiol. Behav. 16*, 431-438.
Elias, M.F., and Eleftheriou, B.E. (1975). *Behav. Res. Meth. Instrumentation 7*, 7-10.
Falconer, D.S. (1960). "Introduction to Quantitative Genetics." Oliver and Boyd, Edinburgh.
Green, E.L., ed. (1966). "Biology of the Laboratory Mouse." McGraw-Hill, New York.
Gurtoo, H.L., Dahams, R., Motycka, L., and Taylor, B.A. (1977). *Fed. Proc. 36*, 939.
Haugen, D.A., Coon, M.J., and Nebert, D.W. (1976). *J. Biol. Chem. 251*, 1817-1827.
Hebeman, R.B., and Aoki, T. (1972). *J. Exp. Med. 136*, 94-111.
Imanishi, T., and Makela, O. (1974). *J. Exp. Med. 140*, 1498-1510.
Kempf, E., Gill, M., Mack, G., Mandel, P. (1978). *C.R. Acad. Sci. (Paris) 238*, 1248-1249.
Lieberman, R., Potter, M., Mushinski, E.B., and Humphrey, W., Jr. (1974a). *Fed. Proc. 33*, 738.
Lieberman, R., Potter, M., Mushinski, E.B., Humphrey, W., Jr., and Rudikoff, S. (1974b). *J. Exp. Med. 139*, 983-1001.
Lieberman, R., Potter, M., and Humphrey, W., Jr. (1975). *Immunogenet. 1*, 529-530.
Lieberman, R., Potter, M., Mushinski, E.B., Humphrey, W., Jr., and Rudikoff, S. (1975). *Immunogenet. 1*, 524.
Lieberman, R., Potter, M., and Humphrey, W., Jr. (1976a). *Fed. Proc. 35*, 549.
Lieberman, R., Potter, M., Humphrey, W., Jr., and Chien, C.C. (1976b). *J. Immunol. 117*, 2105-2111.
Lucas, L.A., and Eleftheriou, B.E. (1975). *Behav. Genet. 5*, 100.

McCarthy, M.M., Taylor, B.A., and Dutton, R.W. (1976). *Fed. Proc.* *35*, 549.
McKenzie, I.F.C. (1975). *Immunogenet.* *1*, 529.
Martelly, I., Bailey, D.W., and DeMaeyer-Guignard, J. (1972). *Ann. Inst. Pasteur* *123*, 835-848.
Mather, K. (1949). "Biometrical Genetics." Chapman and Hall, London.
Meisler, M.H. (1976). *Biochem. Genet.* *14*, 921-932.
Merryman, C.F., and Maurer, P.H. (1975). *Immunogenet.* *1*, 549-559.
Merryman, C.F., Maurer, P.H., and Bailey, D.W. (1972). *J. Immunol.* *108*937-940.
Merryman, C.F., Maurer, P.H., and Stimpfling, J.H. (1975). *Immunogenet.* *2*, 441-448.
Messeri, P., Eleftheriou, B.E., and Oliverio, A. (1975). *Physiol. Behav.* *14*, 53-58.
Mishkin, J.D., Taylor, B.A., and Mellman, W.J. (1976). *Biochem. Genet.* *14*, 635-640.
Mobraaten, L.E., DeMaeyer, E., and DeMaeyer-Guignard, J. (1973). *Transplantation* *16*, 415-420.
Moisset, B. (1977a). *Brain Res.* *121*, 113-120.
Moisset, B. (1977b). *Behav. Genet.* *7*, 78-79.
Morse, H.C., Chused, T., Hartley, J.W., and Taylor, B.A. (1977). *Fed. Proc.* *36*, 1249.
Morse, H.C., Pumphrey, J.G., Potter, M., and Asofski. R. (1976). *J. Immunol.* *117*, 541-547.
Mushinski, E.B., and Potter, M. (1975). *Immunogenet.* *1*, 524-525.
Oliverio. A., ed. (1977). "Genetics, Environment, and Intelligence." Elsevier-North Holland, Amsterdam.
Oliverio, A., and Messeri, P. (1973). *Behav. Biol.* *8*, 771-783.
Oliverio, A., Castellano, C., and Messeri, P. (1972). *J. Comp. Pshysiol. Psychol.* *79*, 459-473.
Oliverio, A., Castellano, C., and Eleftheriou, B.E. (1975a). *Psychopharmacologia* *42*, 219-224.
Oliverio, A., Eleftheriou, B.E., Elias, M.F., and Castellano, C. (1975b). *Psychol. Rep.* *36*, 703-712.
Oliverio, A., Eleftheriou, B.E., and Bailey, D.W. (1973a). *Physiol. Behav.* *11*, 497-501.
Oliverio, A., Eleftheriou, B.E., and Bailey, D.W. (1973b). *Physiol. Behav.* *10*, 893-899.
Potter, M., Finlayson, J.S., Bailey, D.W., Mushinski, E.B., Reamer, B.L., and Walters, J.L. (1973). *Genet. Res.* *22*, 325-328.
Potter, P., Pumphrey, J.G., and Bailey, D.W. (1975). *J. Nat. Cancer Inst.* *54*, 1413-1417.

Reeve, E.C.R., and Waddington, C.H., eds. (1952). "Quantitative Inheritance." H.M.S.O., London.
Riblet, R., Cohn, M., and Weigert, M. (1975). *Immunogenet.* 1, 525-526.
Roderick, T.H., and Schlager, G. (1966). In "Biology of the Laboratory Mouse" (E.L. Greed, ed.), pp. 151-164. McGraw-Hill, New York.
Russel, E.S., Blake, S.L., and McFarland, E.C. (1972). *Biochem. Genet.* 7, 313-330.
Shultz, L.D., and Bailey, D.W. (1975). *Immunogenet.* 1, 570-583.
Shuster, L., Baran, A., and Eleftheriou, B.E. (1975a). *Fed. Proc.* 34, 713.
Shuster, L., Webster, G.W., Yu, G., and Eleftheriou, B.E. (1975b). *Psychopharmacologia* 42, 249-254.
Shuster, L., Yu, G., and Bates, A. (1977). *Psychopharmacologia* 52, 185-190.
Simmel, E.C., Eleftheriou, B.E., Haber, S.B., and Harshfield, G. (1977). *Behav. Genet.* 7, 87.
Snell, G.D., and Stimpfling, J.H. (1966). In "Biology of the Laboratory Mouse" (E.L. Green, ed.), pp. 457-492. McGraw-Hill, New York.
Snell, G.D., Cherry, M., McKenzie, I.F.C., and Bailey, D.W. (1973). *Proc. Nat. Acad. Sci.* 70, 1108-1111.
Stern, R.H., Russel, E.S., and Taylor, B.A. (1976). *Biochem. Genet.* 14, 373-381.
Swank, R.T., and Bailey, D.W. (1973). *Science* 181, 1249-1252.
Taylor, B.A. (1976a). *Genetics* 83, 373-377.
Taylor, B.A. (1976b). *Behav. Genet.* 6, 118.
Taylor, B.A., and Meier, H. (1976). *Genet. Res.* 26, 307-312.
Taylor, B.A., Meier, H., and Myers, D.D. (1971). *Proc. Nat. Acad. Sci.* 68, 3190-3194.
Taylor, B.A., Cherry, M., and Shapiro, L.S. (1973a). *Genetics* 74, s272-s273.
Taylor, B.A., Heiniger, H.J., and Meier, H. (1973b). *Proc. Soc. Exp. Biol. Med.* 143, 629-633.
Taylor, B.A., Meier, H., and Whitten, W.K. (1974). *Genetics* 77, s65.
Taylor, B.A., Bailey, D.W., Cherry, M., Riblet, R., and Weigert, M. (1975a). *Nature* 256, 644-646.
Taylor, B.A., Cherry, M., Bailey, D.W., and Shapiro, L.S. (1975b). *Immunogenet.* 1, 529.
Taylor, B.A., Bedigian, H.G., and Meier, H. (1977). *J. Virology* 23, 106-109.
Thomas, P.E., Hutton, J.J., and Taylor, B.A. (1973). *Genetics* 74, 655-659.

Varnum, D.S., and Stevens, L.C. (1974). *J. Hered.* 65, 91-93.
Watson, J., Riblet, R., and Taylor, B.A. (1977). *J. Immunol.* 118, 2088-2093.
Womack, J.E., Lynes, M.A., and Taylor, B.A. (1975). *Biochem. Genet.* 13, 511-518.

POLYGENE MAPPING: USES AND LIMITATIONS

J. M. Thoday

Department of Genetics
University of Cambridge
Cambridge, England

I. INTRODUCTION

The mapping of components of polygenic systems depends primarily on our ability to detect the effects of allelic differences in a small segment of chromosome independently of the sources of variance arising from segregation and recombination elsewhere and from the inevitable environmental variation.

The ease with which this can be done sufficiently well to meet the purposes of the experimenter depends on the aims of the experimenter, the facilities provided by the genetic system of the organism, our knowledge of that organism, and the particular nature of the variable under study.

An example of the importance of the variable is given in the chapter by Milkman on crossveins in this volume, where the existence of an upper and a lower threshold between which genetic variation gives variable expression means that in a particular cross the number of segregating loci affecting expression is limited.

Except in such a situation success in identifying specific loci depends upon techniques which may allow us to reduce the number of loci segregating, minimise the confusion arising from environmental variance, and sometimes maximise the effect of some particular segregating locus. Given that these can be achieved to a sufficient degree, and given unequivocally classifiable marker genes appropriately linked to some sufficiently effective factor in the polygenic system, that factor can be isolated and mapped in the relevant linkage group.

I should perhaps stress at this point that, despite the

fact that my original paper on this topic (Thoday, 1961) was entitled "The Location of Polygenes", mapping was by no means our primary aim, nor is it the primary aim of many of the workers involved in these studies now. Indeed, it is probably only if one wishes to predict, or explain quantitatively, the effects of selection, taking into account recombination frequency, that accurate linkage maps in the conventional sense might be of great importance. Otherwise, imprecise linkage data may be quite adequate.

The primary aims of such investigations are rather to understand the general architecture of polygenic systems, the nature of polygenes, or the developmental genetics of some important character. For these purposes it may be sufficient that particular segregating units of action can be handled independently. Their precise map positions may, however, be important if we ultimately come to the question of the kind of genetic locus involved in any particular segregation. However, allelism tests, as in the determination of how many may be isoalleles, or other direct methods, as in the demonstration by Frankham *et al.* (1978) that unequal crossing over at the rRNA locus is a source of quantitative genetic variation affecting abdominal bristle number in *Drosophila melanogaster*, can provide conclusions independent of the precision of linkage data.

Mapping in the sense of studying the linkage association between components of polygenic systems and markers is, nevertheless, a main route to isolating and understanding those individual components and their effects, and to supplementing the information available from biometrical analysis (see Jinks, in this volume). I will illustrate the principles with two of the examples from the work, using *Drosophila melanogaster*, with which I have been associated.

First, however, a brief discussion of the two main problems, environmental and background genetic variation, seems to be appropriate.

A. *Environmental Variation*

Basically, the disturbing effects of environmental variation can be minimised in two ways. The first way is to use as controlled environments as possible. The second is to use some device that allows the multiplication up of particular genotypes so that one may compare genotype means, the environmental effects thus being partially averaged out.

The different methods for isolating specific components of polygenic systems have this in common, whether they involve mapping or not. Wehrhahn and Allard's (1965) selection and backcrossing method, Law's (1966) chromosome substitution, recombination and inbreeding method, Swank and Bailey's (1973) recombinant inbred lines (see chapter by Oliverio) and our own *Drosophila* method (see below) all do this in one way or another, and any organism that can be cloned lends itself to such analysis.

B. *Variation in the Genetic Background*

The disturbing effects of segregation at other loci are also minimised by the same methods. With *Drosophila*, however, this may be done more precisely by the use of markers on other chromosomes, or other parts of chromosomes, and by balancer chromosomes. Unwanted recombination can be controlled by the use of males in most *Drosophila* stocks, as well as by balancer chromosomes which, however, may not always function perfectly.

II. FIRST EXAMPLE

For my first example I propose to take the analysis of the chromosome III component of the first polymorphism established by artificial disruptive selection described by Thoday and Boam (1959). The variable was sternopleural chaeta number. The study of the third chromosome was done by Wolstenholme (Wolstenholme and Thoday, 1963).

Thoday and Boam had first identified the chromosome effects using an inbred y bw st marker stock which marks the three major chromosomes. The polymorphism was dependent on heterozygosity in both autosomes. The cross to y bw st both demonstrated this and provided a number of

$$y \frac{bw}{bw} \frac{+}{st}$$

males containing various + third chromosomes which could be maintained by continuous test-crosses to homozygous y bw st females of the inbred stock, thus ensuring control of variation in chromosomes I and II. The degree of control was readily tested by the demonstration that no matter which + chromosome III was involved, the homozygous st flies (white-eyed, since they were also homozygous for the bw marker of chromosome II) were homogeneous between crosses. These crosses showed there were two different classes of + chromosome III, and a possible third, having an intermediate effect on chaeta number relative to the other two, that was almost certainly a

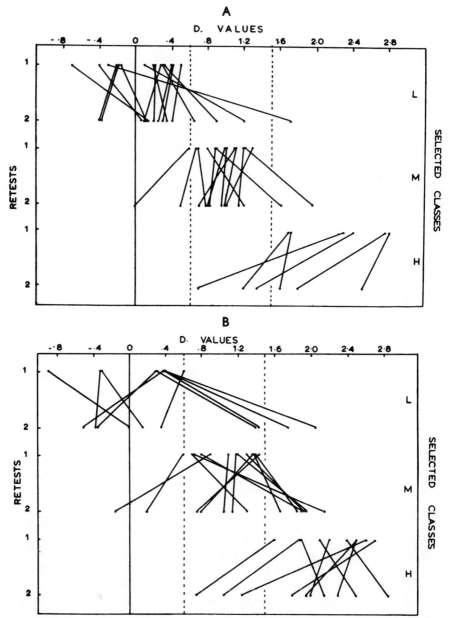

FIGURE 1. Results of two successive progeny tests of cp Sb recombinant chromosomes, selected before progeny testing as High, Medium, and Low. D values are the differences of mean chaeta number distinguishing flies homozygous for the recombinant chromosomes from those homozygous for the recurrent marker standard. A, th st cp + chromosomes; B, + + + Sb chromosomes. (From Wolstenholme and Thoday, 1963).

recombinant.

The analysis of the differences between the two main classes of chromosome III presented considerable difficulty because:

(1) The effects of allelic substitutions at the two loci proved responsible were smaller than those we have studied successfully;

(2) Marker chromosomes suitable for the purpose proved to be in rather heterogeneous backgrounds and some of them contained alleles affecting chaeta number that were not in either class of wild-type chromosome under investigation. Time prevented us from inbreeding each marker stock to reduce the heterogeneity;

(3) A dominant, homozygous lethal marker had to be used at one stage so that recombinants generated had to be test-crossed to a different marker chromosome using such crosses as $+/th\ st\ cp\ Sb \times th\ cp/th\ cp$. This resulted in further background heterogeneity;

(4) One of the high chaeta number chromosomes under test mutated in the course of the breeding programmes.

The effects of this heterogeneity are illustrated in Figure 1. This figure gives the mean chaeta numbers as deviations from those of phenotypically scarlet-pink flies resulting from crosses of the high chaeta number chromosomes as follows:

$$\frac{+\ +\ +\ +}{th\ st\ cp\ Sb}\ \text{females} \times \frac{th\ cp}{th\ cp}\ \text{males}$$

$$\longrightarrow \frac{th\ st\ cp\ +}{th\ cp}\ \text{and}\ \frac{+\ +\ +\ Sb}{th\ cp}\ \text{male progeny.}$$

Several of each of these male progeny are, thereafter, treated separately. These are mated to $st\ p$ as a standard marker.

$$\longrightarrow \frac{th\ st\ cp\ +}{st\ p}\ \text{and}\ \frac{+\ +\ +\ Sb}{st\ p}\ \text{males} \times \frac{st\ p}{st\ p}\ \text{females}$$

The progeny are scored and the males re-mated to $st\ p$ for further progeny tests.

It can be seen from the figure that, though the recombinant chromosomes fall generally into three classes suggesting two effective factors between st and Sb, there is a great deal of misclassification, and any conclusion that only two loci were involved would carry little conviction.

These difficulties were resolved in two ways.

First, an attempt was made to check the results of analysis by synthesis of the original chromosomes. If the conclusions were correct the $th\ st\ cp\ +$ chromosome would comprise three classes: $th\ st\ cp\ H_1\ H_2\ +$, $th\ st\ cp\ L_1\ H_2\ +$, and $th\ st\ cp\ L_1\ L_2\ +$, where H and L are chaeta-number alleles at loci 1 and 2. Correspondingly, the $+\ +\ +\ Sb$ chromosomes should comprise $H_1\ H_2$, $H_1\ L_2$, and $L_1\ L_2$ classes. If we then combine the medium $th\ st\ cp$ and Sb chromosomes in females we should have $th\ st\ cp\ L_1\ H_2\ +/+\ +\ +\ H_1\ L_2\ Sb$ which should generate $H_1\ H_2$ progeny among the $+\ +\ +\ +$ progeny and $L_1\ L_2$ progeny among the $th\ st\ cp\ Sb$ progeny. Further, if no other parts of the chromosomes are variable, then the marker non-recombinants should be homogeneous. Progeny testing the $cp\ Sb$ non-recombinants should confirm this.

The heterozygous females were, therefore, crossed to $th\ cp$ and the male progeny of each marker class were scored and the higher and lower chaeta number males of each class were further progeny tested.

Figure 2 gives the results which confirm the expectations.

The second test involved different markers. The region of the linkage map (Lindsley and Grell, 1967) is: th, 43.2; st, 44.0; cp, 45.3; p, 48.0; cu, 50.0; kar, 51.7; Sb, 58.2.

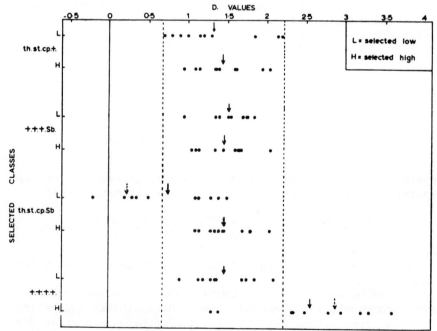

FIGURE 2. *The results of the test for re-synthesis of the chromosomes whose analysis gave the data in Figure 1 (see text). (From Wolstenholme and Thoday, 1963).*

Using the chromosomes $p\ cu$ and $cu\ kar$ we were able to show that there were only two effective factors, one between p and cu, the other between cu and kar, and that the high allele at each locus was fully dominant.

I have given some detail of this investigation since it was the most difficult, yet in the end the most precise of our mapping investigations, and it was what we might call a pure location problem. The corresponding study of the second chromosomes from the polymorphic populations of Thoday and Boam was aided by a pleiotropic lethal interaction between the two loci involved there, that made precise mapping easier. Wolstenholme's study illustrates what can be done in difficult circumstances with an organism in which plenty of markers are available. It also illustrates the problems that can arise when the marker stocks are variable, the critical value of the test synthesis to check the results of analysis, and the degree to which luck in the relation of the loci to useful markers is of importance even with *Drosophila melanogaster*. The $p\ cu\ kar$ region was exceptionally favourable.

III. SECOND EXAMPLE

My other example is the study by Spickett and Thoday (1966) of the second chromosomes from Thoday and Boam's (1961) lines $vg4$ and $vg6$ established by directional selection for sternopleural chaeta number. This illustrates what can be done with markers far apart, and also how enhancement of the effects of a locus by manipulating the genetic background may transform a relatively difficult to a very easy analysis.

The primary aim of this investigation was to determine what genes had been responsible for two accelerated responses in the history of the selection line. All three chromosomes were involved in the difference between the most extreme selection line and one of its ancestral stocks, and there were strong interactions of the other chromosomes with chromosome III.

Thoday, Gibson, and Spickett (1964) showed that all the selected lines had in common a high chaeta number chromosome III containing high alleles at two loci between the markers h and eyg. Their need to use h as a marker which itself raises sternopleural chaeta number substantially, of course, presented a particular problem. Such a problem is, however, soluble by the method first used by Breese and Mather (1957), which involves two rounds of recombination and permits the synthesis of wild-type recombinant chromosomes for test. The markers used are first put into one + chromosome, and then removed with the other, thus allowing the combination of, say, the left region of the chromosome being analysed with the right region of another from a standard inbred line. Each round of recombination must be used to generate a number of recombinants, all thereafter treated separately if the factors are to be mapped.

The interaction between chromosomes II and III meant that it was necessary to analyse chromosome II on two backgrounds, one with a standard chromosome III and the other with the selected III. Both had to be done to demonstrate that the main effect and the interaction depended upon the same genes, but the enhancing background allowed much more clear-cut classification of chromosomes.

The cross used was simple: $H/dp\ cn\ bw$ females were mated to $dp\ cn\ bw$ males to produce several males of each marker recombinant class, thereafter treated separately and backcrossed repeatedly to $dp\ cn\ bw$ females from the stock. Means at each progeny-test generation permit classification of the recombinant chromosomes. Figure 3 shows the results for chromosomes from $vg4$ using an Oregon wild-type stock chromosome I and III. The histograms give the distribution of chromosome means in each recombinant class at first progeny-test and the subsequent progeny-test results are shown below. The $dp\ cn\ bw$ tested against the standard Oregon inbred line proved similar to the Oregon II in effect. Hence, $dp\ cn\ bw$ and $H\ H\ H$ (the chromosomes under test) give the distribution for parental chromosomes. Recombination to the right of cn made no difference. Recombinants between dp and cn were heterogeneous, and on progeny testing, though again there is some misclassification, both marker recombinants fall into two classes, in reciprocal frequencies and with parental means, indicating only one "locus" between dp and cn.

Figure 4 shows the dramatic effect of using the high and enhancing chromosome III background. Classification is unequivocal. The use of an enhancing environment could help in a similar way.

Table I gives the data from these crosses and those involving the other line $vg6$ of Thoday and Boam, that is required for mapping. They are homogeneous and can be combined and give a position for the chaeta number locus at 41.1±1.7 map units. This is fairly precise mapping and, since dp at 13.0 and cn at 57.5 are 44.5 map units apart, shows that very loosely linked markers can be used with success.

Another feature of this example concerns the question of the complexity of the locus identified, which we may consider by asking: "Suppose there were two loci and the recombinants between them were identifiable from their phenotypic effects as an intermediate class of chromosome, how far apart might

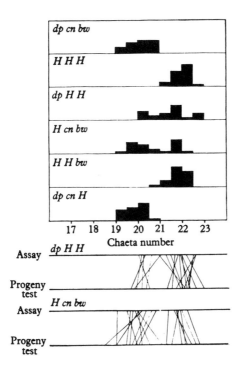

FIGURE 3. *Assay of chromosome II of vg4 against dp cn bw with chromosome III from Oregon. The histograms show the distribution of mean chaeta numbers given by the indicated marker classes of chromosomes. The lower part of the figure shows how repeatable the mean of dp-cn recombinant chromosomes were on retesting by further backcross to the dp cn bw stock. (From Spickett and Thoday, 1966).*

they be and yet recombinants be missed in the sample taken?"

Only the clear-cut segregation on the high chromosome III background can be used for this purpose. The formula for this test is:

$$r_2 = r_1 \chi^2/N+\chi^2,$$

where N is the sample size, r_1 is the map distance between the markers and r_2 that between the two postulated loci. Taking $\chi^2 = 3.841$, the data give 2.038 map units as the maximum for r_2 at the significance level p = 0.05. This again is surprisingly precise, but the formula shows how accuracy with a given amount of labour might be increased with closer markers. Halving r_1 doubles the precision with a given N.

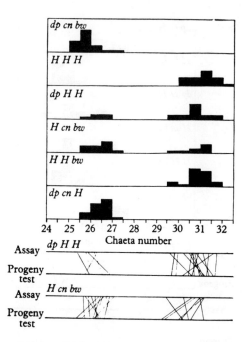

FIGURE 4. Assay of chromosome II of vg4 against dp cn bw with chromosome III from vg4. These histograms show the distribution of mean chaeta number given by the indicated marker classes of chromosomes. The lower part of the figure shows how repeatable the means of dp-cn recombinant chromosomes were on retesting by further backcross to the dp cn bw stock. (From Spickett and Thoday, 1966).

TABLE I. Frequencies of "high" and "low" chaeta-number chromosomes among dp-cn recombinants from the crosses used to assay chromosome II. (From Spickett and Thoday, 1966).

Marker class: Chaeta class:	H cn bw		dp H H	
	High	Low	High	Low
vg4 II (Oregon III)	8	12	11	9
vg6 II (Oregon III)	8	12	14	6
vg4 II (vg4 III)	8	12	15	5
vg6 II (vg6 III)	7	13	12	8
Total	31	49	52	28

$^a\chi^2_{(3)}$ for linkage heterogeneity = 0.94. $P > 0.8$.

IV. MORE COMPLEX PROBLEMS

The examples given above gave relatively simple answers, and in each the alleles under study in the wild-type chromosome had effects detectable in heterozygotes with the marker chromosomes used. In addition, the alleles in each particular chromosome acted in the same direction. There are greater difficulties if the alleles in the marker chromosome are dominant, if more loci are involved, and especially if alleles at different loci in the same chromosome have effects in opposite directions.

A. *Recessives*

If any alleles in the wild-type chromosomes we wish to analyse are completely recessive, they will obviously remain undetected in any programme in which recombinant chromosome means are assessed in backcrosses to marker stocks.

This problem can be resolved if dominant, partially dominant, or co-dominant markers are available, but visible dominants are relatively rare and co-dominant isozyme alleles may involve more work. Alternatively, the recombinant chromosomes can be backcrossed to the wild-type parental stock, assayed and then progeny-tested to the marker stock to identify their marker status. Again more work is involved, but the technique is simple using *Drosophila* stocks that lack male recombination in their autosomes. Another approach is, of course, to use balancers to make marker recombinants homozygous. Again more work is involved, but in addition there are problems arising from failure of balancers especially when balancers for all chromosomes are used in conjunction. Furthermore, there tends to be trouble with genetic background, mutant or otherwise. This points to an essential merit of the repeated backcross programmes which should increase, with each generation, the degree to which any recombinant chromosomes are placed on the genetic background of the marker standard stock being used, and should be especially effective if that stock can be inbred.

It is, of course, always necessary to make homozygotes of chromosomes in which one has isolated particular alleles, if one wishes to assess their dominance relations. At the same time the degree to which the analytic programme has found all the relevant differences can be assessed by determining the extent to which all the parental difference of mean is recovered. Wolstenholme and Thoday (1963) and Spickett and Thoday (1966) did such tests. In the first all the difference was recovered, in the second over 80%.

B. *Larger Numbers of Loci*

When larger numbers of loci are involved, mapping clearly becomes more difficult. Multiple markers can be used as was done by Breese and Mather (1957, 1960), but multiply-marked chromosomes are available in few organisms.

When larger numbers of segregating loci occur between any two available markers, the process of identifying and mapping them is no different in principle, *provided that the alleles are in coupling*, that is to say that all the H alleles are in one parental chromosome and all the L alleles in the other. What is clearly required, however, is an increase in the number of recombinants of each marker reciprocal class that are taken for assay. The success of the analysis must always be a function of that number, whose adequacy can, of course, be assessed by the degree to which consistent conclusions can be derived from the two reciprocal marker recombinants.

C. *Repulsion Linkages*

Gibson and Thoday (1962a) showed that a two locus repulsion linkage could be readily handled with the techniques already described. Their investigation was complicated by lethal interactions between alleles at these two loci such that from the heterozygote $H_1\ L_2/L_1\ H_2$ only the $L_1\ L_2$ recombinant (itself recessive lethal) could be recovered, the $H_1\ H_2$ recombinant being lethal in combination with any of $H_1\ L_2$, $L_1\ H_2$, or (Gibson and Thoday, 1962b) $L_1\ L_2$.

These lethal interactions were of great interest, the position effect involving two loci some 20 map units apart. But the 1962a paper contains a complete demonstration of the two locus situation on the basis of sternopleural chaeta number alone, though more precise mapping could be done using the (apparently pleiotropic) recessive lethality of the $L_1\ L_2$ chromosomes.

If, however, there are mixed coupling and repulsion linkages between two markers, severe problems arise.

D. *Mixed Coupling and Repulsion*

McMillan and Robertson (1974) have drawn attention to the fact that, when there is mixed coupling and repulsion which requires, of course, at least three loci between two markers, the principle of mapping described by Thoday (1961) breaks down. Their paper also contains valuable discussions of the power of the methods for locating polygenes in coupling situations, with assessment of the statistical precision obtainable.

The break down of the principles occurs because these principles depend upon ordering the classes of marker recombinants according to their means and deriving the linkage map from this order as well as from the relative frequencies of the classes.

The model they give is one in which the heterozygote used for analysis is $+ H_1 L_2 H_3 +/a L_1 H_2 L_3 b$ (they point out that the difficulties may not arise if L_2 is nearer one of the markers than both H_1 and H_3). The specific quantitative model they use is as follows, the effect of the a b marker chromosome being taken as zero:

$$a^+ \quad 23cM \quad H_1 \quad 31cM \quad L_2 \quad 7cM \quad H_3 \quad 29cM \quad b^+$$
$$\overline{+1.3-0.8+3.2}$$

Their table 1 gives the resulting conclusions. Three loci are inferred as follows, the quantitative effects and the map positions both being in error.

$$a^+ \quad 23cM \quad H_1 \quad 7cM \quad H_2 \quad 41cM \quad H_3 \quad 29cM \quad b^+$$
$$\overline{+0.5+0.8+2.4}$$

The obvious solution to this problem is to use further markers between a and b, and this should be attempted whenever more than two loci are shown to be involved and appropriate markers are available, but if further usable markers are not available the problem is serious.

Its importance will, however, depend on the aims of the investigation. The errors in the conclusions will profoundly affect their value in explanation of selection responses, especially quantitative explanations that depend on accurate recombination frequencies. However, attempts to study the developmental effects of specific loci will be less affected. Indeed, such attempts might provide clues indicating the error, assuming that the three loci affect the character under study in clearly different ways.

V. EFFECTIVE FACTORS AND LOCATED POLYGENES OR QUANTITATIVE TRAIT LOCI

The components of polygenic systems that can be identified and located by the methods discussed here and elsewhere in this book are strictly to be classed among *effective factors*

in the sense of Mather (see his chapter in this volume). We
identify specific small sections of the linkage groups in
which there are allelic differences affecting the quantitative
character under study.

We will not, of course, identify all the allelic differences for some will have effects too small relative to the
error variance to be detected, but the proportion detected can
be estimated by assessing the proportion of the difference
between parental classes recovered in terms of the effects of
the specific loci.

The question remains whether these effective factors are
particularly complex in the sense that the effective factor
has within it a number of more or less independently acting
genes, or whether a particular analysis has gotten down to a
single gene difference in the same senses as in the study of
major gene differences, a question that must ultimately be
answered if we are to understand the nature of polygenes.
This question is complicated by the complexity of genes in
general.

There is, however, a criterion that may be used to indicate whether single gene differences have been resolved. If
we have identified two or three loci and each is in fact a
complex of several independently acting allelic differences,
it seems highly improbable that most of the sub-loci in any
one effective factor will affect the character in the same
way, a way differing from that produced by most of the loci
in another effective factor. If, therefore, the three loci
affect the character in qualitatively different ways, then it
is probable that they are each to be regarded as single genes.
This is, of course, to use one of the classical definitions
of a gene as a physiological unit.

Spickett (1963) demonstrated that the three effective
factors located by Spickett and Thoday (1966), one of which
provided the second example above, had such specifically different quantitative effects. The three loci affected chaeta-
number in qualitatively very different ways. The same conclusion is to be reached from the fact that two of the loci
showed an interaction effect, but the third interacted with
neither. Thompson's (1975) study of wing vein modifiers
leads to similar conclusions in that different factors affect
different veins or different parts of the same vein.

For these factors, though they may be complex in the
sense that classical genes are complex (and there are indications of such complexity (Thoday, 1973)), we may legitimately
use the term "located polygenes" or "polygenic loci" of
Thompson and Thoday (1974) or the etymologically better term
"Quantitative Trait Loci" (Q.T.L.s) of Geldermann (1975) as
a more precisely defined sub-class of Mather's "effective

factors".

It is these Q.T.L. differences, whose developmental effects can be studied, that give us hope of understanding the nature of polygenes in terms of the kinds of loci that can be involved, and how often differences of structural genes, control genes, repeated sequences, insertions, and so forth, play a part in polygenic systems.

REFERENCES

Breese, E.L., and Mather, K. (1957). *Heredity 11*, 373-395.
Breese, E.L., and Mather, K. (1960). *Heredity 14*, 375-400.
Frankham, R., Briscoe, D.A., and Nurthen, R.K. (1978). *Nature 272*, 80-81.
Geldermann, H. (1975). *Theoret. Appl. Genet. 46*, 319-330.
Gibson, J.B., and Thoday, J.M. (1962a). *Heredity 17*, 1-26.
Gibson, J.B., and Thoday, J.M. (1962b). *Nature 196*, 661-662.
Law, C.N. (1966). *Genetics 53*, 487-498.
Lindsley, D.L., and Grell, E.H. (1967). "Genetic Variations of *Drosophila melanogaster*." Carnegie Institution of Washington Publ. No. 627.
McMillan, I., and Robertson, A. (1974). *Heredity 32*, 349-356.
Spickett, S.G. (1963). *Nature 199*, 870-873.
Spickett, S.G., and Thoday, J.M. (1966). *Genet. Res. 7*, 96-121.
Swank, R.T., and Bailey, D.W. (1973). *Science 181*, 1249-1251.
Thoday, J.M. (1961). *Nature 191*, 368-370.
Thoday, J.M. (1973). *Atti della Academia delli Scienze dell' Instituto di Bologna Anno 261*, 15-26.
Thoday, J.M., and Boam, T.B. (1959). *Heredity 13*, 205-218.
Thoday, J.M., and Boam, T.B. (1961). *Genet. Res. 2*, 161-176.
Thoday, J.M., Gibson, J.B., and Spickett, S.G. (1964). *Genet. Res. 5*, 1-19.
Thompson, J.N., jr. (1975). *Nature 258*, 665-668.
Thompson, J.N., jr., and Thoday, J.M. (1974). *Heredity 33*, 430-437.
Wehrhahn, C., and Allard, R.W. (1965). *Genetics 51*, 109-119.
Wolstenholme, D.R., and Thoday, J.M. (1963). *Heredity 18*, 413-431.

COMPUTER SIMULATION OF THE BREEDING PROGRAM FOR POLYGENE LOCATION

James N. Thompson, jr. and Timothy N. Kaiser

Department of Zoology
University of Oklahoma
Norman, Oklahoma

I. INTRODUCTION

The analysis of quantitative genetic variation can take many forms, depending upon the questions being asked. Other authors have discussed the uses and limitations of biometrical genetics, artificial selection, and other technical approaches to quantitative variation. Although one can estimate gene number and obtain general information about gene effects from these (*e.g.*, Mather and Jinks, 1971; Towey and Jinks, 1977; and extensive discussion in Spiess, 1977), the detailed analysis of specific polygenic effects upon development requires one to be able to isolate and manipulate individual polygenic loci.

In the previous chapter, Thoday discussed polygene mapping and some of the things it can tell us about polygene number and function. Although clearly of great importance, this powerful technique has limitations that are only poorly understood at present (see McMillan and Robertson, 1974; Thompson and Thoday, 1974, for discussions). In this chapter we shall look briefly at one application of computer simulations; we shall describe the design and use of simulations to test the power and limitations of discrimination for a genetic breeding program.

II. COMPUTER SIMULATION OF A BREEDING PROGRAM

A. *The Experimental Location of Polygenic Loci*

Traditional whole chromosome substitution programs, using the balancer chromosomes that characterize *Drosophila* genetic sophistication, enable one to identify the chromosome or chromosomes having significant influence upon the trait being studied. Thus, whole chromosome substitutions allow one to focus attention upon one chromosome set.

A method for intrachromosomal analysis of factors controlling variation in a quantitative trait has been described by Thoday (1961). Thoday's approach integrates the use of

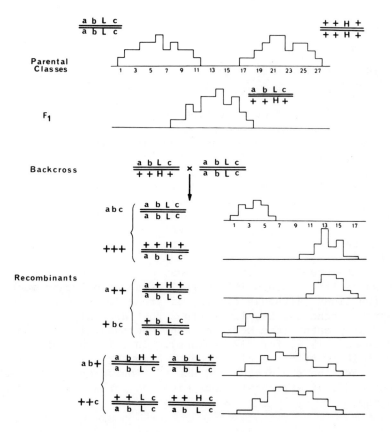

FIGURE 1. *Steps in the hypothetical location of a polygenic locus situated between two recessive flanking markers "b" and "c". (From Thompson, 1977, reprinted with permission).*

chromosome markers to trace sections of selected chromosomes subdivided by recombination (Sax, 1923; Wigan, 1949; Breese and Mather, 1957; Geldermann, 1976; and others) with the method of progeny testing first applied to the analysis of quantitative characters by Johanssen (1909). Assaying the phenotypes associated with marked recombinant segments allows one to determine which segments are important to the expression of the trait. Progeny testing serves the dual role of verifying the phenotypic classification of each recombinant and of identifying the number of relevant linked loci between the marker genes, since the number of loci will be one fewer than the number of phenotypic classes of recombinant chromosomes within a region.

A hypothetical example of polygene isolation is illustrated in Figure 1. An inbred assay stock, marked with the recessive mutants a, b, and c, differs from a selected strain ($+ + +$) in the expression of some target trait. As shown in Figure 1, the $+ + +$ strain is homozygous for a factor (H) that increases the expression of the trait, relative to the L factor in the standard marker line. The H, or High, line might have longer wing veins or more bristles, for example, than the L, or Low, line.

A continuous phenotypic distribution does not necessarily imply the segregation of a large number of polygenes (Thoday and Thompson, 1976). Although we have shown the lines differing by a single factor, it is obvious that in many instances several loci may contribute to the observed phenotypic variation (see McMillan and Robertson, 1974), though this is a reasonable approximation of experiments using close flanking markers. In addition, not all loci are equally effective, and this breeding program allows the most effective factors to be isolated. These more effective loci are, in turn, those that are most likely to be assayable for their developmental effects. The simple locus postulated here has a significant influence upon the phenotype and is located between known markers, the loci outside this narrow region having been eliminated by chromosome substitution and recombination prior to this stage in the program. The variance in the two parental strains is due to uncontrolled environmental variation.

The F_1 progeny of a cross between the two parental lines are intermediate in expression of the trait. They are phenotypically wild type for the three recessive markers, and heterozygous for the polygenic locus. A simple testcross of heterozygous females to homozygous standard assay strain males will result in a series of recombinant chromosomes that can be recognized phenotypically: $a\ b\ c$, $+ + +$, $a + +$, $+ b\ c$, and so forth. Only the single recombinants are shown in this figure.

The next step is to assay each recombinant for its expression of the quantitative trait. The resulting distributions are shown at the right of Figure 1. Let us look at the $a + +$ recombinant class first. If the effective region is between "b" and "c", all $a + +$ recombinant chromosomes will carry the H allele ($a + H +$). From the backcross they will also carry the $a\ b\ L\ c$ chromosome, and they will have intermediate expression of the character (a mean of about 13 units in this example). The $+ b\ c$ chromosomes will all carry the L allele, and the backcross progeny will be homozygous for L allele, resulting in a phenotypic distribution like that of the standard (parental) Low expression strain. The recombinants between "b" and "c", however, can be of two types: those with the break occurring between "b" and the effective region (*i.e.*, the postulated polygenic locus) and those with the break occurring between the effective region and "c". Thus, the $a\ b +$ recombinant class is composed of two chromosome types, producing a wide distribution of quantitative phenotypes.

The question that remains is: How many of the chromosomes in the heterogeneous distribution represent recombination to the left of the polygenic locus and how many represent recombination to the right. This can be established by progeny-testing, that is, by test-crossing and reassaying each recombinant chromosome for several generations. The key to polygene location, therefore, is to sample progeny containing known recombinants in each progeny-test generation and average their phenotypes, thereby statistically eliminating the random environmental effects that obscure the genetic contribution to the phenotype.

B. *The Computer Model of the Breeding Program*

The computer simulation of the polygene location technique has, as its primary objective, the analysis of hypothetical location experiments designed to test the power of this experimental breeding program. In each run, the question is simply whether random environmental effects can be eliminated sufficiently to allow one to discriminate polygenic effects of a given magnitude.

A simple matrix arrangement was used to represent the chromosome makeup and genetic recombination. Heterozygous diploid "chromosome sections" were created by establishing two 1×100 matrices. Each position in each matrix was given a base value of 0.1, providing a background phenotypic value of 20 in the homozygous standard diploid. A hypothetical polygenic locus was, then, represented by entering the magnitude of its phenotypic effect at a particular position, 2 to 99, in one of the matrices. The position selected, thus, determines the linkage with respect to the flanking markers (located at positions "1" and "100"). Multiple loci can be represented, and both cis and trans arrangements can be modelled easily. In our initial studies, a position of "50" has been used, and polygene magnitudes have ranged from 1 to 3.

Recombinant chromosomes are generated by "breaking and rejoining" the matrices at a point determined by a random number generator. Then, the values at each position in the "recombined" matrix are summed, thereby classifying each chromosome into one of two types by whether it carries the polygenic locus or not. Phenotypic distributions are drawn from the sample of recombinant chromosomes after random environmental variation is introduced using random normal deviates (Snedecor, 1956) with a variance set by the experimenter.

Progeny-testing is readily modelled by independently determining the phenotypic effect of a recombinant chromosome (genotype plus random environmental effect) a certain number of times and then averaging the phenotypes to plot a single point per generation. This can be done for any number of recombinant chromosomes from the original distribution and can be continued for any number of progeny-test generations.

III. RESULTS AND FUTURE APPLICATIONS

Many experiments have shown that, for certain types of quantitative traits, a large proportion of the variation can be accounted for by the segregation of a small number of polygenic loci (Thompson and Thoday, 1974). In fact, even a single segregating locus affected by a small amount of random environmental variation can produce a phenotypic distribution essentially indistinguishable from that produced by segregation at hundreds of loci (see demonstration in Thoday and Thompson, 1976). This is confirmed by our present simulations of which two are reproduced in Figures 2 and 3. A single polygenic locus is segregating in each of these. In both, the locus has a relative phenotypic effect of 2.5 with a standard level of environmental variation. This environmental variation completely masks individual genotypes producing an essentially continuous phenotypic distribution, even though the sample consists of approximately half recombinant and half non-recombinant chromosomes.

The simulations shown here do, however, illustrate one of the most powerful aspects of the breeding program -- the discriminating ability provided by the progeny-testing. In

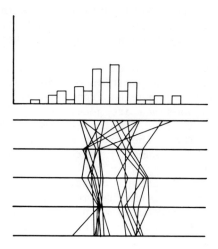

FIGURE 2. *A simulation of 50 recombinant chromosomes and progeny-testing of 15 of these for four generations. In each line, ten individuals are scored and averaged to obtain the point plotted for each progeny-test generation.*

Figure 2 the phenotypes of ten offspring are averaged to obtain an estimate of the genotypic effect of each recombinant chromosome for four consecutive generations. By the second generation of progeny-testing, the two sets of chromosomes are

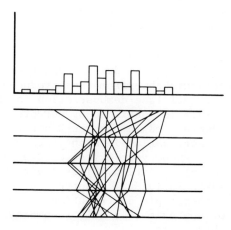

FIGURE 3. *A simulation identical to that shown in figure 2, except only five are averaged to plot the points in each progeny-test generation.*

unambiguously separated and an estimate of map position can be calculated.

One of the parameters we have been studying is the size of progeny-test samples. In Figure 3, although the same genetic parameters determine the distribution, only five offspring are measured and averaged to obtain the point plotted in each generation. The environmental effects are so large, however, that the two chromosome classes are not distinguishable.

Using such simulations as these, one is able to estimate the minimum relative magnitude of polygenic effect that can be detected under various conditions. For example, with five individuals averaged per chromosome each generation, a gene magnitude to environmental variation ratio of 1:0.6 can be resolved in five or fewer generations of progeny-testing, while with ten individuals averaged, a gene having an effect approximately 10% smaller can be discerned with no difficulty.

As pointed out by McMillan and Robertson (1974) in their computer analyses, complications such as linkage, in which no crossovers happen to occur between two closely-linked factors, will lead one to underestimate the number of loci involved and to overestimate the magnitude of their phenotypic effects. These sources of error are particularly severe when dealing with chromosomes in which H and L factors are linked in cis configurations, or where inequality of gene effects or other biases occur (see, for example, Park, 1977a, 1977b). Only a small beginning has been made towards an understanding of the factors that can influence our ability to identify individual polygenic loci. But even with the potential sources of error discussed here, the polygene location technique clearly provides a powerful method for studying the major contributors to quantitative genetic variation.

ACKNOWLEDGMENTS

We would like to thank Mr. Steven Graves for his assistance in designing this program. This is part of a study of the frequency and magnitude of polygene mutation supported by DHEW-NIEHS grant 5-R01-ES01439-03.

REFERENCES

Breese, E.L., and Mather, K. (1957). *Heredity 11*, 373-395.
Geldermann, H. (1976). *Theoret. Appl. Genet. 47*, 1-4.

Johanssen, W. (1909). "Elemente der exakten Erblichkeitslehre." Fischer, Jena.
McMillan, I., and Robertson, A. (1974). *Heredity 32*, 349-356.
Mather, K., and Jinks, J.L. (1971). "Biometrical Genetics." Chapman and Hall, London.
Park, Y.C. (1977a). *Theoret. Appl. Genet. 50*, 153-161.
Park, Y.C. (1977b). *Theoret. Appl. Genet. 50*, 163-172.
Sax, K. (1923). *Genetics 8*, 552-560.
Snedecor, G.W. (1956). "Statistical Methods." Iowa State College Press, Ames, Iowa.
Spiess, E.B. (1977). "Genes in Populations." Wiley, New York.
Thoday, J.M. (1961). *Nature 191*, 368-370.
Thoday, J.M., and Thompson, J.N., jr. (1976). *Genetica 46*, 335-344.
Thompson, J.N., jr., and Thoday, J.M. (1974). *Heredity 33*, 430-437.
Towey, P., and Jinks, J.L. (1977). *Heredity 39*, 399-410.
Wigan, L.G. (1949). *Heredity 3*, 53-66.

POLYGENIC INFLUENCES UPON DEVELOPMENT
IN A MODEL CHARACTER

James N. Thompson, jr.

Department of Zoology
University of Oklahoma
Norman, Oklahoma

I. INTRODUCTION

Essentially all metrical characters appear to be influenced by polygenic segregations. Indeed, as Lewontin (1974) has pointed out, when experiments are properly designed, there appears to be only one major failure to modify any of the hundreds of morphological, behavioral, physiological, or cytological traits that have been studied in *Drosophila*. This "failure" was, however, quite instructive. It involved an attempt by Maynard Smith and Sondhi (1960) to produce a left-biased asymmetry in the pattern of bristles on the heads of *Drosophila subobscura* carrying the mutant *ocelliless*. The lack of response to selection suggested that there was no left-to-right developmental gradient information involved in the formation of the dorsal bristle and ocelli pattern on the *Drosophila* head. The "failure" to produce a particular response implied the absence of genes with a particular effect and, thus, contributed a significant insight into the genetic control of the development of a morphological pattern.

Artificial selection, with accompanying genetic analysis provided by such specialized techniques as polygene mapping (Thoday, 1961), may in fact be one of the most powerful tools we have for identifying the fragile connections between genes and the developmental processes they control. This chapter will discuss the types of developmental questions that are most readily answered using the techniques of quantitative genetics and will illustrate some of them using the system of polygenes that modify wing vein length and pattern in the species in which polygenic analysis is most readily accomplished, *Drosophila melanogaster*.

II. DEVELOPMENTAL QUESTIONS AND THE CHOICE OF A MODEL EXPERIMENTAL SYSTEM

In studies of simple organisms, it is commonly possible to isolate discrete developmental processes. But most eukaryotes are much too complex for such a simplistic approach, particularly when one wants to investigate the development of quantitative traits or the developmental significance of polygenic modifiers of these traits. Wallace (1968) was quite correct when he wrote that an understanding of the precise action of genes cannot come directly from an analysis of phenotype frequencies. Some important questions about development can, however, be answered using the powerful techniques of quantitative genetics and the subtle changes in phenotype that are, by definition, the core of this branch of genetics.

Studies of the polygenic modifiers of wing vein length and pattern began with an interesting paradox. Although the length and position of *Drosophila* wing veins is essentially invariant in nature, all sampled populations contain polygenes that modify the expression of all vein mutants that have been studied. If the polygenes are not expressed in the adult phenotype of wild flies, why are they present in the gene pool and what, if any, functions do they serve? Such a problem can be divided roughly into two components: "How many polygenes influence the expression of the phenotype?" and "What can one conclude about the ways in which they modify developmental activities?" (Thompson, 1975a).

Although estimates of polygene number can be made in a variety of ways (Thoday, 1961; Milkman, 1970; Mather and Jinks, 1971; and others), one must think very carefully about the types of developmental questions that are appropriate for a polygenic system. Studies in quantitative genetics are primarily concerned with gene frequencies and with the distribution of phenotypes that their segregation produces. But phenotype frequencies can tell us nothing about primary gene action in the sense of "What is the gene product and how does it function in development?" On the other hand, the variation and responses of quantitative traits to selection and to environmental stresses can tell us a lot about the ways in which developmental processes interact to shape the phenotype. To be more specific, it can show us the relationship between the general phenotypic measures of gene expression such as dominance, penetrance, and expressivity, and it can clarify the

TABLE I. *Classification of mutants of Drosophila melanogaster arranged according to the veins they affect (from Thompson, 1974a).*

Extra veins	Shorten veins[a]				Number in class
	L_2	L_3	L_4	L_5	
X	--	--	--	--	38
--	X	--	--	--	7
--	--	X	--	--	2
--	--	--	X	--	10
--	--	--	--	X	8
--	--	--	X	X	7
--	?	X	?	X	1
--	X	?	X	X	3
--	?	X	X	X	1
--	X	X	X	X	5

[a]*Many genes do not appear to shorten all veins, except in some genetic backgrounds, suggesting thresholds in expression. Those which might fit into the "all-vein shortened" category are listed separately, but questionable veins are indicated with a question mark.*

role pleiotropy plays in determining correlations among characters. Perhaps most interesting at present, however, it can help us to understand the complexity of patterns, such as wing vein arrangement, and to understand the way minor gene effects influence the interaction of developmental sequences that result in the final phenotype.

The pattern of veins in *Drosophila melanogaster* is an ideal model system. First, the pattern is simple, consisting of four primary longitudinal veins (the anterior wing margin is vein L_1 and the other veins are numbered sequentially) and two crossveins. A large number of mutations affecting vein directly (or as a pleiotropic effect upon cuticle formation) are known, and many have been studied extensively (Table I). Some affect vein formation in general by adding or eliminating vein material in all parts of the wing, while others are region-specific, affecting only one vein or one area of the wing. Second, as noted before, vein mutants show a great amount of variation when outcrossed to wild-type lines, and they respond readily to artificial selection. Third, the application of quantitative techniques requires that measurements of the phenotype be easily made and accurate. Wing and vein lengths can be measured easily with a microscope eyepiece graticule, and the ratio of vein length to wing length gives a simple correction for fly size. Finally, the histological development of the wing is well-known (*e.g.*, Waddington, 1940). The wing develops from an imaginal disc which can be dissected and manipulated throughout development. In addition, special cell autonomous markers enable sophisticated clonal analyses to be carried out, and biochemical studies and tissue culture can be used to follow the development of normal and mutant wings (see Garcia-Bellido, 1972; Schneiderman, 1976; Whittle, 1976). Thus, the histological and cellular effects of selected genetic backgrounds and of isolated polygenic loci are potentially accessible to study, though little of this work has yet been done.

III. RESPONSE TO SELECTION

The study of patterns can be approached in a variety of ways. Waddington (1973), for example, stressed that the "causal structure of a pattern ... can be successfully investigated only when we can find ways of producing controlled changes in it." One way of producing controlled changes is by artificial selection, in which the phenotype is modified in a predetermined direction by small steps.

Selection responses can show the rate of change of a phenotype, correlated responses, and genetic limits to change. One of the most powerful applications of artificial selection is seen, however, in the analysis of physiological limits. An example of this is found in studies of *veinlet* (Figure 1), a mutant in which all longitudinal veins are shortened.

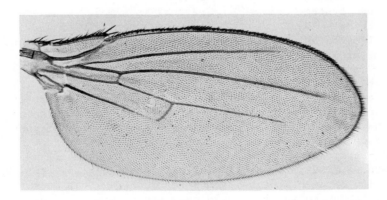

FIGURE 1. *The mutant veinlet in Drosophila melanogaster.*

A. *Separability of Polygene Effects*

The length of each vein in the recessive mutant *veinlet* varies in expression after outcrossing to wild-type flies from a natural population. The question one can ask is to what extent do the same polygenic modifiers act upon different veins? In other words, do the polygenes act upon the expression of the mutant *veinlet* in a general way, or do they act upon the components of the *veinlet* phenotype by modifying separate vein lengths? This can be answered in at least two ways. First, one could select for opposite responses of the longitudinal veins in a single selection line. Second, one could assay the chromosomal contributions to *veinlet* expression to determine whether different chromosomes had different contributions to expression of the L_2, L_3, and L_4 veins. Both types of experiments have been done.

Correlation between veins could be due either to a shared physiological process or to linkage. Selection for opposite responses of different veins in a single line, which one might call "dissectional selection", is a design that allows one to dissect phenotypic correlations. For example, Davies and Workman (1971; Davies, 1971) used such a program to establish that the determinants of variation in abdominal and sternopleural bristle numbers were to a large extent independent and distinct. In order to look for separability of the responses of individual veins, two *veinlet* selection lines were set up: one in which the L_2 and L_3 veins were selected to be longer and the L_4 was selected to be shorter, and one in which only the L_3 was lengthened while both the L_2 and the L_4 were shortened. Both lines responded successfully, though there appeared to be some interdependence between the L_2 and L_3 veins (Thompson, 1975b). This degree of interdependence is not surprising, however, since both are in the anterior half of the wing which is physiologically distinguishable from the posterior half containing the L_4 and L_5 veins (Garcia-Bellido *et al.*, 1973).

Assays of heterozygous whole chromosome contributions to vein length confirmed the fact that L_2, L_3, and L_4 vein polygenes were distinct from one another (Table II). The major effects for lengthening the L_2 vein, for example, are associated with chromosome II, while those for the L_3 and L_4 are associated with chromosome III. The separability of the L_3 and L_4 effects had already been demonstrated in the "dissectional selection" lines, thus suggesting that the L_3 and L_4 were simply linked on chromosome III rather than being identical. This was confirmed when an L_4 vein-specific polygenic locus $(PL(2)L_4a$; Thompson, 1975b) was mapped to position 72.5 using the polygene location method of Thoday (1961).

TABLE II. Summary of the effects of heterozygous whole autosomes (II^s and III^s) from representative veinlet selection lines. (From Thompson, 1975b).[a]

Selection line	Vein	II^s	III^s	Total[b]
ve Long-III	L_2	+0.052***	+0.016***	+0.068
ve Long-III	L_3	+0.005	+0.023***	+0.028
ve Long-III	L_4	+0.006	+0.041***	+0.047
ve Short-II	L_2	+0.016***	−0.042***	−0.026
ve Short-II	L_3	−0.031***	+0.004	−0.027
ve Short-II	L_4	−0.096***	−0.004	−0.100
ve Short-IV	L_2	+0.012	−0.014**	−0.002
ve Short-IV	L_3	−0.012***	+0.004*	−0.008
ve Short-IV	L_4	−0.061***	+0.027	−0.034

[a] levels of significance: *, $0.05 > p > 0.01$; **, $0.01 > p > 0.001$; ***, $p < 0.001$.

[b] Total includes interchromosomal interactions not shown in this abstracted table.

B. *Range of Phenotypic Expression*

Although selection lines of a large variety of mutants have shown that the amount of vein material can be modified dramatically, one of the most interesting applications of selection lines involves their use in determining the range and limits of phenotypic expression. The recessive mutant *plexus* adds vein fragments to all parts of the wing. By selection, the amount of extra vein material can be increased (Figure 2) and can be reduced until the wing is indistinguishable from wild-type. At levels of high expression vein fragments are present in rather specific positions on the wing. Moreover, it appears that with increasing mutant expression, vein material is added in a fairly predictable manner. By analyzing the position of veins and fragments, the order in which they appear, and the rate at which they become larger, a three-dimensional picture of the probability of vein formation can be drawn, in which the wing surface is two dimensional and the probability curves are the third dimension. A similar method of representing developmental potentials has been used by others (*e.g.*, Maynard Smith and Sondhi, 1961; Carlson, 1970).

This analysis of the range of phenotypic expression can lead in two directions: to a more precise description of the development of vein pattern and to an analysis of the evolution of vein pattern.

Let us look first at the view of development that this information from selection lines provides. Remember that there are two types of vein mutants, those that affect vein expression throughout the wing and those with region-specific (vein specific) effects. Waddington (1940) described the histological development of several extra-vein and vein-gap mutants and found that the mutant effects were first expressed in the presence or absence of basal processes between the cell layers of 19-28 hour old pupae. He showed that the origin of vein material depends upon the forming and maintaining of spaces

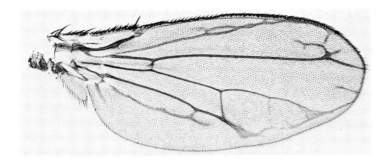

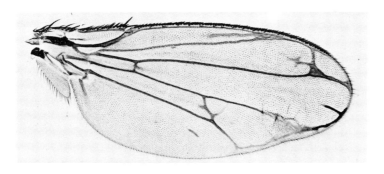

FIGURE 2. Pattern of extra veins in high selected lines of the mutant plexus of Drosophila melanogaster. The "new" longitudinal vein in the third posterior cell (above) is analogous to a similar vein in related families of Diptera. Its physiological independence from the other extra vein fragments is shown by its sensitivity to lower temperatures, in that it is eliminated when flies from the high plexus line are grown at $18°C$, rather than the normal culture temperature of $25°C$.

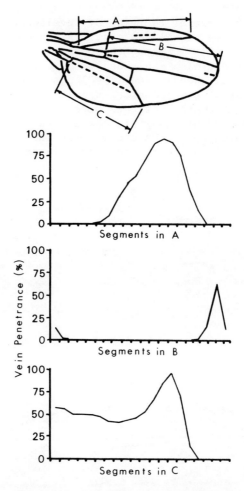

FIGURE 3. Quantified pattern profiles for the fragments that appear in the marginal cell (a), the first posterior cell (b), and the third posterior cell (c). The region of the wing which is represented in each profile is illustrated at the top. (From Thompson, 1974a).

between the two cell layers of the pupal wing, for in the final stage of wing development, extra chitin is deposited in all areas in which the two layers are not touching. Where basal processes form, the two cell layers come into contact and form a central membrane. Where basal processes have not formed, spaces remain between the layers and extra chitin is later deposited there. Mutants with extra veins appear to have the basal processes inhibited in specific parts of the wing, while in the vein-gap mutants the spaces are erased when basal processes appear soon after the normal vein canals have formed. Thus, *plexus*-like extra vein mutants and vein-gap mutants such as *veinlet* can be seen as opposite ends of a continuum of vein expressions.

On the other hand, individual veins and fragments form in particular regions of the wing, and the level of their capacity to respond is also variable. One example (Figure 2) is the sensitivity to temperature shown by the "new" longitudinal vein in the third posterior cell which is almost completely eliminated in flies grown at cool temperatures. Though put in different ways by several workers (*e.g.*, House, 1953; Scharloo, 1962; Carlson, 1970; Thompson, 1974a), it is thought that the developmental potentials of individual veins might be viewed as the expression of a prepattern and that the vein-specific mutants affect components of that prepattern. The analysis of quantitative expression of veins, insofar as it allows one to detect gradual changes in phenotype, enables us to map the contours of this prepattern. An example is shown in Figure 3. Here the likelihood of formation of vein material in 5% units along the length of a specific vein or vein fragment yields a map of the vein-forming-ability of regions of the wing. Indeed, the delicacy of quantitative variation in monitoring the degrees of expression provides an exceptional means of detailing the underlying processes of development.

The second direction in which analysis of the range and limits of phenotypic expression can lead is into a discussion of evolution of phenotypes. One elegant example is provided by the work of Sondhi (1962) who subjected a population of *Drosophila subobscura* which carried the mutant *ocelli-less* to selection for an increase in the number of ocelli and in the number of certain bristles in the ocellar region of the head. In one line a bristle occurred that was constant in position and orientation and which was similar in both respects to a pair of macrochaetes in the family Aulacigastridae, a family of Diptera closely related to the Drosophilidae. Such observations are noteworthy, not in that they necessarily support suggested taxonomic relationships, but in that they indicate that there exist in wild-type individuals of a species a number of developmental potentials which are not expressed in the adult phenotype. These "subthreshold" developmental potentials are, nevertheless, retained within the genetic and developmental organization and can be revealed by certain environmental or genetic influences such as selection.

The "subthreshold" pattern of extra veins and fragments is basically the same in *plexus* and *net* (Thompson, 1974a). In addition, though degrees of plexation vary, parts of the pattern are also observed in other selection lines and in most other extra-vein mutants (see Lindsley and Grell, 1967). If the pattern were a non-specific side effect of the primary mutational change, such similarities would probably be very limited. Finally, veins or vein modifications which appear to be analogous to the subthreshold veins described in these selection lines (Thompson, 1974a) have also been described by others in *D. melanogaster* (Goldschmidt, 1945), in other *Drosophila* species (*e.g.* Gersh, 1972), and in numerous related and ancestral Diptera (*c.f.* Oldroyd, 1970; Comstock, 1918).

It might be expected, then, that as the subthreshold pattern appears to be reasonably stable, it could reflect regions in which ancestral veins had once occurred, but which are still represented in the developmental architecture of the wing either because they are intimately involved in the present wing development or because natural selection to remove them becomes ineffective when their expression is decreased. Numerous parallels between the subthreshold veins and the venation in related Diptera support this hypothesis. The observed correlations with related Dipteran patterns imply that one way in which the present *Drosophila* pattern might have evolved is by the selection of polygenic modifiers or weak mutant alleles that gradually erased the ancestral veins from the evolving wing, leaving only the series of subthreshold potentials (the unexpressed prepattern).

One of the questions that evolutionary biologists ask is "How is a new structure or function originated through evolutionary change?" Although the elimination of veins from the ancestral wing is a relatively simple structural change, the insights we have of the genetic and developmental interrelationships that may have produced these changes can contribute, in some degree, to an understanding of the problems involved. Whatever structure or function evolves, it must be modified from something that is already present in the developmental repertoire of the organism, be it an undifferentiated cell or a complex organ. The living organism is a highly complex system of processes and any change must maintain that organization. Major mutations are generally harsh in their effects upon an organism, as judged from the lower relative fitness of mutant individuals. Thus, the genetic changes that will normally be necessary to evolve a new structure or function will be those of small effect, *i.e.* polygenes producing minor modifications of an existing structure or function. Indeed, the small changes associated with polygenes and their interactions may be the only way in which the fine-tuning of the living organism can be maintained as it adapts to new environments. That gene interactions can be important is shown, for example, by Gibson (1970) who found that bristle number selection lines at different temperatures primarily altered the positive interactions between second and third chromosomes.

IV. THE NATURE AND NUMBER OF POLYGENES

As sometimes interpreted (*e.g.*, Vetta, 1976), the classical definition of polygenes with its implied prediction of polygene number is inconsistent with cytogenetic estimates of the number of genes in *Drosophila* (Thompson, 1975a). Although polygenes may affect any number of steps in protein synthesis and development (see Synthesis by Thompson and Thoday in this volume), there is no evidence that most polygenes are in any way different, except in magnitude of expression, from other genes.

A. *General Versus Mutant-specific Polygene Effects*

The argument that polygene number must be extremely large often takes the form that, if character a is affected by a number of polygenes and so are characters b, c, d, and so on, the number of accumulated polygenes must be very large. This may be true, but it does not mean that the number of genes affecting a specific trait is extremely large.

Misunderstanding of polygene number is often based upon the assumption that polygene effects are specific to the character being measured, rather than simply affecting some underlying developmental process that might be shared by a number of phenotypic characters. For example, for those polygenes that are observed to modify the expression of a mutant, such as *veinlet* or *plexus*, polygenes would be thought to be mutant-specific modifiers of mutant expression. Several lines of evidence suggest that this is generally not true.

One piece of evidence comes from the analysis of polygenes affecting vein length. Several of the mutants studied by Thompson (1973, 1974b) affected the same vein. Thus, one could ask whether the same polygenes, or more precisely the same selected whole chromosomes, affected the length of a particular vein the same way in different mutants. Similarly, one could ask whether chromosomes carrying polygenes that increased the amount of vein in extra-vein mutants also increased vein length in vein-gap mutants. The results are summarized in Table III. It can be seen that, in general, a chromosome has similar phenotypic effects when substituted into the background of a different mutant. Where the chromosome increased the length of the L_4 vein or the amount of vein

TABLE III. *Summary of chromosome substitutions, carrying selected polygenes, from one mutant background into another. (From Thompson, 1975a).*[c]

Vein affected by polygenes	Chromosome substituted[a]	First mutant[b] (source)	Second mutant	Effect in first mutant	Effect in second mutant
L_2	II	ve L-2	ri	+	+
L_2	II	ve L-3	ri	+	+
L_2	II	ve S-2	ri	+	none
L_2	II	ri L	ve	+	+
L_2	III	tg^c S	shv	−	−
L_3	II	ri L	ve	none	none
L_3	II	ri S	ve	none	none
L_4	II	ve S-2	ci	−	−
L_4	II	ve S-4	ci	−	−
L_4	II	ri L	ci	none	+
L_4	II	ri L	ve	none	none
L_4	III	shv L	ci	+	+
L_4	III	shv S	ci	−	−
L_4	III	ve S-4	ci	−	−
L_4	III	ve S-4	shv	−	−
L_4	III	px L	shv	+	+
L_4	III	px S	shv	−	−
L_4	III	net L	shv	+	+
L_4	III	net S	shv	−	−

[a] *Chromosome extracted from the selection line (without carrying a venation mutant) and substituted into a control line of the second mutant.*

[b] *The original selection line. L, longer veins or extra vein material; S, shorter veins or fewer vein fragments; numbers refer to tests of parallel selection lines of the same mutant.*

[c] *Mutant symbols: tg^c (telegraph of Carlson) and ri (radius incompletus) shorten vein L_2; ci (cubitus interruptus) shortens vein L_4; ve (veinlet) and shv (short vein) shorten all veins; net (net) and px (plexus) have fragments of extra venation.*

material in one mutant, for example, it did so in the other mutants as well. Where the chromosome had no detectable effect, its action upon the expression of the other mutant was random or not detectable. The only exceptions involved the line of the L_2 vein mutant *radius incompletus* (*ri*) selected for longer veins and might be due to linkage or interactions.

These results were confirmed by the identification and location of genes within the effective chromosome. One specific example has been studied in detail (Thompson, 1975b; Thompson and Thoday, 1975). A single locus on the second chromosome was found to shorten the L_4 vein and to act as a modifier of *veinlet* and *cubitus interruptus*. Indeed, its effects were even limited to a specific part of the L_4 vein, a subterminal region distal to the crossvein.

The absence of mutant-specific effects of vein length polygenes suggests that they act directly upon the vein-forming ability of particular veins (L_2, L_3, and so forth) rather than as control genes or modifiers of the expression of specific mutants or loci, though some polygenes may theoretically act in that way. Since the same genes can affect several different mutants or traits, it can readily be seen that polygene number, as such, is not necessarily inconsistent with cytogenetic evidence since the number of genes needed to explain the known genetic variation is not as large as earlier interpretations had led us to expect.

B. *Polygenes as Isoalleles*

The general absence of mutant-specific phenotypic effects leads naturally to the hypothesis that polygenes may be isoalleles of loci also known through mutant alleles of larger effect. Isoalleles or suppressed "major mutant alleles" have been found in selection lines by Milkman (1970), Scharloo (1970), Waddington (1961), Bateman (1959) and others.

On the other hand, Waddington *et al.* (1957) failed to find isoalleles of *cubitus interruptus Wallace* during selection in an isogenic line. In addition, Thompson (1975c) showed that one cannot necessarily predict the effect which particular isoalleles will have upon phenotypic variation. In this study isoalleles of *cubitus interruptus* were substituted into the genetic background of *veinlet* (a mutant that is developmentally similar enough to share a proportion of their polygenic modifiers, *c.f.* Table III). Surprisingly, the isoalleles had no detectable effect upon the length of the L_4 vein (Table IV).

With these cautions in mind, it is still reasonable to assume that a large proportion of the polygenes affecting the phenotypic variation in a trait are isoalleles of loci that might also be known through major visible or lethal mutant alleles.

TABLE IV. Mean relative vein lengths of the L_4 veins in control and experimental veinlet flies.[a]

Isoallele of ci	Experimental		Control	
	Females	Males	Females	Males
$+^3$	0.584	0.565	0.565	0.559
$+^5$	0.570	0.556	0.563	0.555
$+^{01}$	0.547	0.540	0.538	0.535
$+^{02}$	0.569	0.550	0.564	0.551

[a] (From Thompson, 1975c). Experimental flies are heterozygous for the isoallele and for an inbred standard chromosome; only heterozygous effects are measured. Vein lengths are expressed as the mean sine transformed ratio of vein length to total wing length. $N = 180$ in each of the four sets of chromosomes tested; 15 replicate cultures were sampled in each analysis. Only the difference between $+^3$ and the control chromosome was significant; the $+^3$ chromosome increased, not decreased, the L_4 vein length.

C. Character Definition

The evidence that polygenes are isoalleles or mutations of small effect acting upon the developmental processes underlying a character helps us understand the apparent inconsistencies among estimates of polygene number affecting different types of traits.

It is intuitively clear that not all phenotypes are equally complex. Wing vein length or sternopleural bristle number are reasonably simple quantitative traits, while body weight and height are more complex. Spickett (1963) showed that variation in sternopleural bristle number could be accounted for by polygenic loci that affected cell number, cell size, cell division rate, and the distribution of bristle precursor materials. The cell development loci could, in turn, also affect other quantitative traits, including body size (though suppressor or enhancer loci sometimes compensate for changes not directly selected for, see Spickett, 1963).

On the other hand, quantitative variation in body weight can be produced by variation in genes affecting the efficiency of nutrient uptake, rate of digestion, metabolic rate, bone length, muscle mass, endocrine function, cell size and rate of cell division, and many other contributory processes. Indeed, many quantitative geneticists look at these traits individually, but sometimes fail to realize that the trait they define for study is a component of other, more encompassing or complex, quantitative traits. Thus, the choice of trait can lead to quite different estimates of gene number, making the analysis of gene number *per se* an exercise of limited usefulness (Thompson, 1976). The more useful approach, therefore, is one in which primary attention is given to the nature and developmental action of polygenic loci, rather than to assessing sheer number of factors.

V. QUANTITATIVE VARIATION AS A COMPLEMENT TO MOSAIC ANALYSIS IN DEVELOPMENTAL GENETICS

Because of the extreme phenotypes that can be generated and the range of intermediate expressions possible, selection lines can be useful complements to standard analytical tools of developmental genetics such as mosaic analysis. A brief example will illustrate their possible uses.

Although we now have a reasonably good idea of the ways in which cells become gradually differentiated during development, the genetic control of spatial relationships within an embryo are much more poorly understood. The wing vein system of *Drosophila melanogaster* provides an excellent experimental system for studying pattern formation, and some aspects of the problem have been discussed in earlier sections of this chapter. One basic question which has not been mentioned, however, is whether the veins (or the prepattern of vein-forming ability discussed earlier) develops as a consequence of oriented cell divisions during the growth of the wing blade, or whether it is imposed upon the wing at a later stage in development by a tissue-level mechanism such as diffusion. This question has been a focus of activity in my laboratory and to answer it we have implemented mosaic analysis in selected vein mutant lines (J.N. Thompson, J.V. Toney, and G.B. Schaefer, unpublished).

The first question we asked was whether vein pattern was determined by, or independent of, cell lineage. To answer this we looked at the relationship between the position of the veins and the anterior-posterior compartment boundary in the wing (Figure 4), as described by Garcia-Bellido *et al.* (1973). The boundary between anterior and posterior wing cells is a straight line just anterior to the L_4 vein; mixing of cells between these groups does not occur, as monitored by large clones of cells marked with the autonomous cell hair (trichome) marker *multiple wing hair* (*mwh*). Thus, the question is whether the shape of the compartment boundary varies in selection lines of a mutant like *plexus*, in which the L_4 vein is

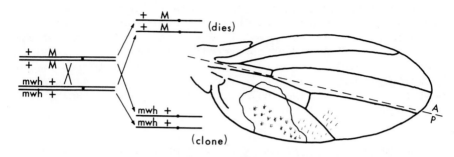

FIGURE 4. *The anterior-posterior (A-P) boundary in the Drosophila wing, as identified by the behavior of large clones produced by the "Minute technique" of Garcia-Bellido et al. (1973). Heterozygous Minute (M) flies develop more slowly than normal, due to a decrease in the rate of protein synthesis. When + M/mwh + cells are irradiated, somatic recombination yields cells homozygous for the Minute mutation which die, and cells homozygous for the recessive cell marker mwh. The mwh cells lacking Minute grow at a normal rate (faster than the surrounding Minute-carrying cells) and produce a large clone of marked cells. It is found that this clone, whether anterior or posterior, does not cross a compartmental boundary line just anterior to the L_4 vein. (Diagram from Schaefer and Thompson, 1977).*

bent posteriorly when the amount of venation is increased (see Figure 2). The answer was: no characteristic change in compartment boundary occurred in the selection line, implying that vein formation is, at least to some extent, independent

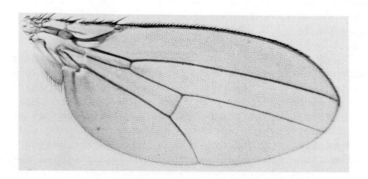

FIGURE 5. *Selected telegraph of Carlson (tg^C) in which the expression of the L_2 vein has been totally eliminated.*

of cell lineage.

Indeed, this conclusion is confirmed by a simple consideration of the behavior of clones in the *Minute* technique (Figure 4; Schaefer and Thompson, 1977). If vein formation were determined early in development, then the overgrowth of a large clone would be expected to disrupt the vein pattern. No such disruption is found; the vein pattern in large clones is identical to that in a normal wing.

This leads one to question what mechanism might be responsible for determining vein arrangement. Again, selected lines can be useful. One hypothesis is that the vein pattern is determined by interacting diffusion gradients formed during the pupal stage of wing blade formation. A prediction from this hypothesis is that the elimination of a vein or part of a vein should cause compensating movements of veins due to the changed interactions of the diffusion gradients. A suitable selection line is shown in Figure 5; the L_2 vein of this selected line of tg^C is completely absent. The question then becomes: "Are there changes in the distances between veins or other alterations of vein pattern in response to the loss of one component?" To answer this, we simply measured the distances between veins, vein lengths, angles, and areas in this selection line and in control flies and found that the distance did indeed increase in the selected tg^C wing, thus supporting the hypothesis that the vein pattern is determined by diffusion gradients late in development.

This developmental analysis of selection lines is quite distant from the questions of polygene number and function, and though a powerful complement to the standard techniques of developmental genetics, it cannot stand on its own. It should illustrate, however, the richness of information that can be obtained from the careful selection of experimental systems and should show that quantitative systems, in spite of their innate complexity, have much to offer geneticists in many fields.

ACKNOWLEDGMENTS

I would like to thank Mr. Joe V. Toney and Mr. G. Bradley Schaefer for their help in developing the information summarized in the last section.

REFERENCES

Bateman, K.G. (1959). *J. Genet. 56*, 341-352.
Carlson, J.H. (1970). *Ohio J. Sci. 70*, 365-371.
Comstock, J.H. (1918). "The Wings of Insects." Comstock Publ. Co., Ithaca, New York.
Davies, R.W (1971). *Genetics 69*, 363-375.
Davies, R.W., and Workman, P.L. (1971). *Genetics 69*, 353-361.
Garcia-Bellido, A. (1972). In "The Biology of Imaginal Disks" (H. Ursprung and R. Nöthinger, eds.), pp. 59-91. Springer-Verlag, Berlin.
Garcia-Bellido, A., Ripoll, P., and Morata, G. (1973). *Nature New Biol. 245*, 251-253.
Gersh, E.S. (1972). *Genetics 72*, 531-536.
Gibson, J.B. (1970). *Heredity 25*, 591-607.
Goldschmidt, R. (1945). *Univ. California Publ. Zool. 49*, 291-550.
House, V.L. (1953). *Genetics 38*, 309-327.
Lewontin, R.L. (1974). "The Genetic Basis of Evolutionary Change." Columbia Univ. Press, New York.
Lindsley, D.L., and Grell, E.H. (1967). "Genetic Variations of *Drosophila melanogaster*." Carnegie Inst. of Washington Publ. No. 627.
Mather, K., and Jinks, J.L. (1971). "Biometrical Genetics." Chapman and Hall, London.
Maynard Smith, J., and Sondhi, K.C. (1960). *Genetics 45*, 1039-1050.
Maynard Smith, J., and Sondhi, K.C. (1961). *J. Embryol. Exp. Morph. 9*, 661-672.
Milkman, R.D. (1970). *Adv. Genet. 15*, 55-114.
Oldroyd, H. (1970). "Handbooks for the Identification of British Insects: Diptera. I. Introduction and Key to Families." Roy. Entomol. Soc., London.
Schaefer, G.B., and Thompson, J.N., jr. (1977). *Bios 48*, 103-109.
Scharloo, W. (1962). *Arch. Neerl. Zool. 14*, 431-512.
Scharloo, W. (1970). *Genetics 65*, 681-691.
Schneiderman, H.A. (1976). In "Insect Development" (P.A. Lawrence, ed.), pp. 3-34. Blackwell, Oxford.
Sondhi, K.C. (1962). *Evolution 16*, 186-191.
Spickett, S.G. (1963). *Nature 199*, 870-873.
Thoday, J.M. (1961). *Nature 191*, 368-370.
Thompson, J.N., jr. (1973). *Genet. Res. 22*, 211-215.
Thompson, J.N., jr. (1974a). *Heredity 33*, 389-401.
Thompson, J.N., jr. (1974b). *Heredity 33*, 373-387.
Thompson, J.N., jr. (1975a). *Nature 258*, 665-668.

Thompson, J.N., jr. (1975b). *Genetics 81*, 387-402.
Thompson, J.N., jr. (1975c). *Heredity 35*, 401-406.
Thompson, J.N., jr. (1976). *Nature 261*, 525-526.
Thompson, J.N., jr., and Thoday, J.M. (1975). *Genet. Res. 26*, 149-162.
Vetta, A. (1976). *Nature 261*, 525-526.
Waddington, C.H. (1940). *J. Genet. 41*, 75-139.
Waddington, C.H. (1961). *Adv. Genet. 10*, 257-293.
Waddington, C.H. (1973). In "Developmental Systems: Insects. Volume 2" (S.J. Counce and C.H. Waddington, eds.), pp. 499-535. Academic Press, New York.
Waddington, C.H., Graber, H., and Woolf, B. (1957). *J. Genet. 55*, 246-250.
Wallace, B. (1968). "Topics in Population Genetics." Norton and Co., New York.
Whittle, J.R.S. (1976). In "Insect Development" (P.A. Lawrence, ed.), pp. 118-131. Blackwell, Oxford.

GENES AFFECTING QUANTITATIVE ASPECTS OF PHYSIOLOGY IN RODENTS

John G. M. Shire

Institute of Genetics
University of Glasgow
Glasgow, Scotland

I. INTRODUCTION

Many characters show apparently continuous phenotypic variation. This chapter describes some physiological systems in which the individual genes contributing to such variation within the range of normality have been identified. This has been done in two ways. Sometimes the phenotypic variation has been controlled by repeated redefinition of the character, with each definition more specific than its predecessor. In other cases, the potential number of loci has been restricted, by minimizing the genetic divergence between the stocks being compared. In some investigations both these approaches have been used.

II. PHENOTYPIC REFINEMENT

A. *Redefinition of the Character*

Lewis Dahl and his colleagues (Knudsen *et al.*, 1970) produced, by selective breeding, lines of rats which differed in their responses to a high-salt diet. Rats of the S line rapidly developed severe hypertension whilst the blood-pressure of those of the R line did not increase. The F_1 rats produced by crossing these lines had an intermediate response. A wide range of phenotypes were found in both the F_2 and backcross generations, with no evidence of assortment into distinct classes. John Rapp looked at the level of corticosteroids in peripheral plasma (Rapp and Dahl, 1971), since these hormones regulate salt metabolism. He found significant differences between the S and R lines in the mean concentration of both 11-hydroxy deoxycorticosterone (corticosterone) and 18-hydroxy deoxycorticosterone (18OH DOC), although there was considerable overlap between the strains when measurements on individual rats were compared. The relative proportions of the two steroids were, however, quite distinct for the two lines. Studies on the production of these steroids by homogenates of adrenal tissue (Rapp and Dahl, 1972) showed that there was no overlap between the values for the R and S lines in the rate of production of either steroid from their common precursor, deoxycorticosterone (DOC). The mean rate of production of these steroids, by 11- or 18-hydroxylation of DOC, by adrenal tissue from F_1 rats was intermediate between those of the parental lines. The range of values for individual F_1 rats overlapped both parents.

When the phenotype was defined as the ratio of 18-hydroxylation to 11-hydroxylation the distribution for the S, R and F_1 rats showed no overlap. The values for the F_2 rats fell into three distinct classes, whilst those for the rats of the backcrosses fell into two clear-cut classes. The alleles at this locus affecting steroid synthesis appear to define different forms of the cytochrome P450 common to the 18- and 11-hydroxylase reactions (Rapp and Dahl, 1976). Measurements on a sample of F_2 rats showed that segregation at this locus accounted for some, but not all, of the difference in blood pressure between S and R rats on the high salt diet. The exact proportion of the selection response accounted for by this locus cannot be determined exactly from the data of Rapp and Dahl (1972), because of the small sample sizes, combined with the use of the mean rather than the mode of the F_2 families.

The adrenal glands of male mice of the CBA strain weigh almost twice as much as those of males of the A strain. The distributions of adrenal weight for the F_1, F_2, and backcross hybrids between these strains were continuous (Badr and Spickett, 1971). However, when the volumes of the component parts of the gland were studied, the data suggested that only a few loci were involved (Shire, 1970). Variation at one locus, with heterozygotes having an intermediate phenotype, accounted for the observed difference in the size of the cortex. Variation at two loci, with dominance for the CBA alleles, was sufficient to account for the observed differences in the volume of the medulla.

The total production of adrenal steroids by A and CBA mice did not differ, even when expressed per unit body weight as in

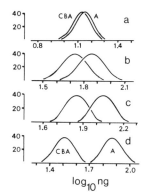

FIGURE 1. *Adrenal steroid production in male mice of the A and CBA strains. a) total steroids per g body weight, b) total steroids per mg adrenal, c) total steroids per mg cortical tissue, d) corticosterone per mg cortical tissue. (Redrawn from Spickett et al., 1967, with permission).*

Figure 1. When steroid production was expressed per mg of adrenal tissue, the strains differed significantly, but their distributions overlapped considerably (Spickett *et al.*, 1967). This overlap was reduced when the phenotype was redefined in terms of the amount of steroidogenic (cortical) tissue within the adrenals. When the production of a particular hormone, corticosterone, was expressed in this way the phenotype showed a discontinuous distribution. Using the ratio of corticosterone to steroids which were not 11-hydroxylated, Badr and Spickett (1965) were able to show clear-cut segregations in the F_2 and backcross to CBA generations. The dominant allele at this locus controlling hormonal output by the adrenal is present in mice of the A strain.

B. *Environmental Alterations*

The genetic factors underlying phenotypic variation can often be made more amenable to study by altering environmental conditions (see Milkman, 1979). The levels of corticosterone in the plasma of mice of four strains were similar when the undisturbed animals were compared, but were markedly different after the mice had been exposed to a standardized stress (Levine and Treiman, 1964). The difference between C57BL and DBA mice in the circulating levels of corticosterone following stress were found to be the result of differences in both the rate of corticosterone production by the adrenals and the rate of its degradation by the liver (Doering et al, 1972; Lindberg et al., 1972). The ease with which the differences between C57BL and DBA mice can be demonstrated depends on the diet that they are fed (Shire, 1978). The difference between the strains is greatest when their diet is low in fat.

Exposure to cold, or to the hormones released during stress, revealed striking differences between DBA, CBA, and C57BL mice in the magnitude and rapidity with which an enzyme activity in the adrenal medulla increased. The resting, steady-state, levels of the enzyme did not, however, differ very much between these strains (Ciaranello et al., 1972; Ciaranello, 1978).

III. CONTROL OF GENETIC DIVERSITY

A. *Comparison of Related Lines*

The DBA and C57BL inbred strains, started 20 years apart from very different mouse stocks, differ at more loci than inbred strains which have recent ancestors in common. Half the genome of both the CBA and C3H strains was derived from DBA, and the other half from BALB (Staats, 1966). Thus, these strains should differ at fewer loci than do DBA and C57BL, or DBA and BALB. This should simplify the analysis of the lively behavior of the F_1 mice produced by crossing docile CBA and DBA mice.

Sublines separated after considerable inbreeding of the parental strains should differ at very few loci (Bailey, 1978). Thus, quantitative characters which do differ between such sublines should have relatively simple genetics. The level of catalase differs twofold not only between strains, but also among sublines of the C57BL strain. Most C57BL sublines have low levels of catalase, but three closely-related sublines have higher levels. The difference between the two groups of sublines was shown to be due to variation at a single locus, *Ce*. Variation at this locus produces differences in the steady-state level of catalase by altering the rate at which this enzyme is degraded within liver cells (Paigen, 1971). Variants at the structural locus (*Cs*) for catalase, which is on a different chromosome, cause differences in the level of catalase activity not only in the liver, but also in other tissues.

Ciaranello (1978; Ciaranello et al., 1974a, 1974b) showed that whilst mice of the J subline of BALB had high levels of three of the enzymes which synthesize epinephrine, mice of the N subline had low levels (Table I). Measurements on hybrids

TABLE I. *Levels of tyrosine hydroxylase (TH), dopamine β-hydroxylase (DBH), and phenylethanolamine N-methyltransferase (PNMT) in the adrenals of BALB mice.*[a]

Subline	TH	DBH	PNMT
/cJ	8.9±1.2	30.2±2.4	0.37±0.1
/cN	4.5±0.5	17.3±1.2	0.19±0.1

[a] Activities expressed as nanomoles of product formed per hour per adrenal pair. (From Ciaranello et al., 1974b, with permission.)

between the sublines suggested that the differences in the levels of all three enzyme activities were controlled by a single, segregating, element. As with catalase, the difference affects the rate of degradation or inactivation of the enzymes, rather than the rate of synthesis of their primary amino acid sequence.

There are many congenic pairs of inbred strains. In such stocks a particular variant from one inbred strain has been placed, by repeated backcrossing, on the genetic background of another inbred strain. Sibs are then mated within each congenic stock to maintain, or produce, homozygosity at all loci. Differences between such congenic lines have been found in such characters as the size of the immune response to particular antigens (Gasser and Shreffler, 1972), and in the responsiveness of target tissues to sex-steroids (Gregorova, 1977) and corticosteroids (Bonner and Slavkin, 1975; Pla et al, 1976; Goldman et al., 1977). These differences have been attributed to variation at loci close to the locus at which the congenic strains are known to differ. However, even after ten generations of inbreeding, considerable heterozygosity will still exist in the variant line. This heterozygosity will not be restricted to loci linked to the variant whose segregation is being controlled (Fisher, 1949; Falconer, 1961). Inbreeding such a congenic line will make unlinked heterozygous regions homozygous, but with equal chances of fixing variants derived from either of the two original strains (see section IV).

B. *Major Mutants Can Be Useful*

Individuals with a mutant phenotype are often phenotypically more variable than wild-type individuals. In some cases this may be due to the overt phenotypic effects of genes whose expression could not be detected in non-mutant individuals (Rendel, 1979).

The Oligosyndactyly (*Os*) mutation in the house mouse is a recessive lethal. It has dominant effects on the morphology of the foot and the development and function of the kidney (Falconer, 1964; Stewart, 1978). The severity of the diabetes insipidus which stems from the renal changes depends on the genetic background of the stock in which the mutation segregates. Studies on backcross mice, produced by crossing lines with severe and mild diabetes insipidus, have revealed genetic variation affecting tubules (Stewart, 1971), and the quantity, but not the affinity, of the kidney receptors for antidiuretic hormone (Dousa and Valtin, 1974; Stewart, 1978).

Major mutants have been profitably combined with the use of related sublines. The diabetes mutation (*db*) produces severe, often lethal, diabetes mellitus in the KsJ subline of C57BL, but only mild diabetes in the 6J subline (Hummel *et al.*, 1972). When the unlinked mutation obese (*ob*) was placed on these two backgrounds, a similar difference in the severity of its effects was found (Coleman and Hummel, 1973). Before these experiments there was no evidence for any difference between these sublines in the control of carbohydrate metabolism. Using a linked marker gene, Gunnarsson (1975) was able to maintain a mild expression of *db* typical of the 6J subline even after four backcrosses to the KsJ subline. This observation, together with the high degree of relatedness of the sublines, suggests that a few loci are involved, and that they might be located fairly easily.

IV. IMPORTANCE OF POST-SEGREGATIONAL STUDIES

A. *Crossing the Original Lines*

Considerable insight into the genetic and physiological organization of a character can be gained by crossing two lines which differ quantitatively in that character. Studies on F_1 crosses will reveal new phenotypic combinations when loci with different forms of inheritance affect a character. Adrenal weight and testis weight were negatively correlated over the A and CBA strains, but not when reciprocal F_1 mice were studied (Shire, 1976). C57BL mice have larger stores of esterified cholesterol, the precursor of all steroid hormones, in both testis and adrenal than are found in DBA mice (Doering *et al.*, 1972). This correlation did not hold for C57BL × DBA mice. In both cases, attention was directed towards processes specific to each tissue, and away from processes common to both steroid producing organs (Shire, 1976).

It is important to measure individuals from hybrid generations in which segregation of the various genetic factors can occur. Investigations of the relationship between specific phenotypic characters are usually more fruitful than experiments in which a wide range of different characters are measured (Stewart and Elston, 1973). Rapp's (1972) studies on the F_2 of the rat lines selected for altered blood pressure showed that the difference in the proportions of 11- and 18-hydroxylation segregated as a unit, and was not due to separable differences in the two hydroxylases. In contrast, a positive correlation between cholesterol ester stores and corticosterone production in C57BL and DBA mice and their F_1 hybrids broke down when backcross mice were measured (Doering et al., 1973), showing more than one locus was involved. One of the loci involved is $ald-2$, which controls the size of the cholesterol ester stores of adult mice. AKR and DBA mice are phenotypically indistinguishable, so far as the stress response, development, and precursor stores of the adrenal glands are concerned. Observations on backcross mice showed that whilst both strains differed from C57BL at a single locus controlling cholesterol ester levels, the loci ($ald-1$ and $ald-2$) were different (Doering et al., 1973; Shire, 1978).

Measurements on hybrids between sublines drew Roland Ciaranello's attention to a mechanism common to three enzymes (Ciaranello et al., 1974a). Studies on F_2 or backcross mice could show whether that biochemical difference is the cause of the difference in aggressive behavior found between the same sublines (Ciaranello et al., 1974b).

Mice of the C57BL/10 and B10.A congenic strains differ at the H-2 histocompatibility locus, and also in the size of various tissues which are target-organs for male sex hormones (Gregorova et al., 1977). Measurements on F_2 mice bred by crossing the congenic lines showed the weights of these organs were affected by independently segregating factors, as well as by the H-2 chromosomal region. Similar studies should be made on congenic lines which differ at the H-2 locus and also in susceptibility to cleft-palate induced by the injection of cortisone (Bonner and Slavkin, 1975; Goldman et al., 1977). One study on backcrosses between a susceptible strain (A) and a resistant one (C57BL) suggests that the H-2 region does contain an important determinant of susceptibility (Biddle and Fraser, 1977).

B. *Derived Lines for Difficult Characters*

Some characters cannot be measured on single individuals either because the measuring methods are not sensitive enough, or because environmental sources of variation are large. These characters can sometimes be analyzed by progeny-testing individuals of segregating generations (Thoday, 1966). Such an approach has shown that most of the differences between CBA and RAP mice in the rate of sodium excretion by the kidney was controlled by variation at a single locus (Stewart and Mowbray, 1972).

There are characters, such as the slope of a dose-response curve or the effect of dietary alterations on liver enzymes, which require mutually exclusive measurements to be made on the same individuals. Such characters can be investigated with recombinant inbred lines. These sets of lines are produced by crossing two standard inbred strains, and then inbreeding samples of the F_2 or backcross generations (Oliverio, 1979).

Ovulation rarely occurs in response to a light-flash in C57BL/By females, but can easily be induced in BALB/c By and F_1 mice (Figure 2). Recombinant lines derived from a cross between the two strains fell into two clearcut classes, each resembling one parental strain (Eleftheriou and Kristal, 1974). The BALB allele at the H-18 histocompatibility locus was present in all the responsive recombinant lines. Females of the B6G. $H-18^c$ line, in which the H-18 allele typical of BALB is carried on a C57BL background, were also found to ovulate in response to environmental stimuli. Thus, the locus controlling susceptibility may be on chromosome 4. Similar

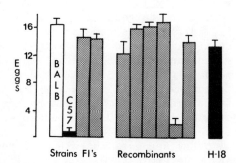

FIGURE 2. *The number of eggs ovulated per female in response to a standard flash of light. The mean values are shown for the BALB/c By and C57BL/By strains, their reciprocal F_1's, six recombinant lines, and a line congenic with C57BL. (Drawn from data in Eleftheriou and Kristal, 1974, with permission.)*

studies have suggested that the difference between C57BL and BALB mice in the resting levels of corticosterone is controlled by variation at two loci (Eleftheriou and Bailey, 1972). A third locus seems to control the difference between these strains in the binding of corticosterone in the hypothalamic region of the brain (Eleftheriou, 1974).

V. PROSPECTS

Studies of carefully defined phenotypes in laboratory populations will go a long way towards identifying the factors which underlie variation within the range of normality in physiological systems. Once identified, such genetic factors can be located, not only to particular chromosomal positions but also in relation to particular DNA sequences, whether or not they code for amino acid sequences (Gilbert, 1978). As we learn more about the mode of action of these genes they will be found to include variants at both structural and "control" loci. Some variants will prove to be isoalleles at loci originally identified through alleles with pathological effects, whilst others will map to previously unknown loci.

All the examples that I have used in this chapter concern laboratory populations, all of which are to some degree inbred. Even so, individual laboratory populations of Sprague-Dawley rats are polymorphic for adrenal hydroxylase activity (Rapp and Dahl, 1971). It will be interesting to discover the forms and extent of genetic variation affecting homeostatic systems in natural populations under different ecological conditions (Berry, 1977). Such studies will require a reliable set of laboratory stocks for progeny-testing and some, at least, of the phenotypes to be defined in terms of protein structure.

REFERENCES

Badr, F.M., and Spickett, S.G. (1965). *Nature 205*, 1088-1090.
Badr, F.M., and Spickett, S.G. (1971). *J. Endocrin. 49*, 105-111.
Bailey, D.W. (1978). *In* "Workshop on the Origins of Inbred Mice" (S. Morris, ed.). Academic Press, New York.
Berry, R.J. (1977). "Inheritance and Natural History." Collins, London.
Biddle, F.G., and Fraser, F.C. (1977). *Genetics 85*, 289-302.
Bonner, J.J., and Slavkin, H.C. (1975). *Immunogenetics 2*, 213-218.

Ciaranello. R.D. (1978). *In* "Genetic Variation in Hormone Systems" (J.G.M. Shire, ed.), Ch. 12. CRC Press, Cleveland.

Ciaranello, R.D., Dornbusch, J.N., and Barchas, J.D. (1972). *Molec. Pharmacol. 8*, 511-520.

Ciaranello, R.D., Hoffman, H.J., Shire, J.G.M., and Axelrod, J. (1974a). *J. Biol. Chem. 249*, 4528-4536.

Ciaranello, R.D., Lipsky, A., and Axelrod, J. (1974b). *Proc. Natl. Acad. Sci., U.S.A. 71*, 3006-3008.

Coleman, D.L., and Hummel, K.P. (1973). *Diabetologia 9*, 287-293.

Doering, C.H., Shire, J.G.M., Kessler, S., and Clayton, R.B. (1972). *Endocrinology 90*, 93-101.

Doering, C.H., Shire, J.G.M., Kessler, S., and Clayton, R.B. (1973). *Biochem. Genet. 8*, 101-111.

Dousa, T.P., and Valtin, H. (1974). *J. Clin. Invest. 54*, 753-762.

Eleftheriou, B.E. (1974). *Brain Res. 69*, 77-82.

Eleftheriou, B.E., and Bailey, D.W. (1972). *J. Endocrin. 55*, 415-420.

Eleftheriou, B.E., and Kristal, M.B. (1974). *J. Reprod. Fertil. 38*, 41-47.

Falconer, D.S. (1961). "Introduction to Quantitative Genetics." Oliver and Boyd, Edinburgh.

Falconer, D.S., Latyszewski, M., and Isaacson, J.H. (1964). *Genet. Res. 5*, 473-488.

Fisher, R.A. (1949). "The Theory of Inbreeding." Oliver and Boyd, Edinburgh.

Gasser, D.L., and Shreffler, D.C. (1972). *Nature, New Biol. 235*, 155-156.

Gilbert, W. (1978). *Nature 271*, 501.

Goldman, A.S., Katsumata, M., Yaffe, S.J., and Gasser, D.L. (1977). *Nature 265*, 643-645.

Gregorova, S., Ivanyi, P., Simonova, D., and Mickova, M. (1977). *Immunogenetics 4*, 301-313.

Gunnarsson, R. (1975). *Diabetologia 11*, 431-438.

Hummel, K.P., Coleman, D.L., and Lane, P.W. (1972). *Biochem. Genet. 7*, 1-13.

Knudsen, K.D., Dahl, L.K., Thompson, K., Iwai, J., Heine, M., and Leitl., G. (1970). *J. Exp. Med. 132*, 976-1000.

Levine, S., and Treiman, D.M. (1964). *Endocrinology 75*, 142-144.

Lindberg, M., Shire, J.G.M., Doering, C.H., Kessler, S., and Clayton, R.B. (1972). *Endocrinology 90*, 81-92.

Milkman, R.D. (1979). *In* "Quantitative Genetic Variation" (J.N. Thompson and J.M. Thoday, eds.). Academic Press, New York.

Oliverio, A. (1979). *In* "Quantitative Genetic Variation" (J.N. Thompson and J.M. Thoday, eds.). Academic Press, New York.
Paigen, K. (1971). *In* "Enzyme Synthesis and Degradation in Mammalian Systems" (M. Rechcigl, ed.), pp. 1-46. University Park Press, Baltimore.
Pla, M., Zakany, J., and Fachet, J. (1976). *Fol. Biol. Praha* 22, 49-52.
Rapp, J.P., and Dahl, L.K. (1971). *Endocrinology* 88, 52-65.
Rapp, J.P., and Dahl, L.K. (1972). *Endocrinology* 90, 1435-1446.
Rapp, J.P., and Dahl, L.K. (1976). *Biochemistry* 15, 1235-1242.
Rendel, J.M. (1979). *In* "Quantitative Genetic Variation" (J.N. Thompson and J.M. Thoday, eds.). Academic Press, New York.
Shire, J.G.M. (1970). *J. Endocrin.* 48, 419-431.
Shire, J.G.M. (1976). *Biol. Rev.* 51, 105-141.
Shire, J.G.M. (1978). *In* "Genetic Variation in Hormone Systems" (J.G.M. Shire, ed.). Ch. 3. CRC Press, Cleveland.
Spickett, S.G., Shire, J.G.M., and Stewart, J. (1967). *Mem. Soc. Endocrin.* 15, 271-288.
Staats, J. (1966). *In* "The Biology of the Laboratory Mouse" (E.L. Green, ed.), pp. 1-9. McGraw-Hill, New York.
Stewart, A.D. (1978). *In* "Genetic Variation in Hormone Systems" (J.G.M. Shire, ed.). Ch. 8. CRC Press, Cleveland.
Stewart, J. (1971). *Pflüg. Arch.* 327, 1-15.
Stewart, J., and Elston, R.C. (1973). *Genetics* 73, 675-693.
Stewart, J., and Mowbray, S. (1972). *Genet. Res.* 19, 61-72.
Thoday, J.M. (1966). *In* "Proceedings of the Third International Congress of Human Genetics" (J.F. Crow and J.V. Neel, eds.), pp. 339-350. Johns Hopkins Press, Baltimore.

CYTOLOGICAL MARKERS AND QUANTITATIVE VARIATION IN WHEAT

C. N. Law and M. D. Gale

Plant Breeding Institute
Cambridge, England

I. ANEUPLOIDS AND THE CHROMOSOME CLASSIFICATION OF WHEAT

The hexaploid bread wheat, *Triticum aestivum* (2n = 6× = 42), exhibits disomic inheritance and is composed of three distinct sets of 14 chromosomes. The chromosomal sets or genomes are related to each other in that they are derived from diploid species which have participated in the evolution of wheat and which, in their turn, have evolved from a common ancestral diploid. This triplication of genetic information enables wheat to tolerate the loss or addition of entire chromosomes without disastrous consequences on the phenotype. A wide range of chromosomally deficient or aneuploid lines involving each of the 21 wheat chromosomes have been obtained. These are all viable and can readily be maintained.

The aneuploid lines of wheat have been used to determine the origin or genomic status of each wheat chromosome as well as the genetical relationships that occur between them. Thus, it has been possible to show that apart from the chromosomal classification into genomes, the chromosomes of wheat can be separated into seven homoeologous groups, each composed of three chromosomes, one from each genome, and each having a high degree of genetical similarity (Sears, 1954, 1966). This chromosome classification in terms of origin and homeology is shown in Table I and defines, albeit rather broadly, the genetical relationships that exist between wheat chromosomes.

Besides the aneuploids produced by the loss of whole chromosomes, other lines are available which have incomplete chromosomes resulting from the misdivision of chromosomes at meiosis. One of the most useful of these misdivision products is

TABLE I. *The chromosome classification of hexaploid wheat, Triticum aestivum.*

		Genomes		
		A	B	D
Homoeologous groups	1	1A	1B	1D
	2	2A	2B	2D
	3	3A	3B	3D
	4	4A	4B	4D
	5	5A	5B	5D
	6	6A	6B	6D
	7	7A	7B	7D

the telocentric chromosome in which one of the arms of a chromosome has been lost. Since the centromeres of wheat chromosomes are median or sub-median, then telocentric chromosomes can readily be distinguished from complete chromosomes in dividing somatic and meiotic cells. It is this ease of recognition which enables telocentric chromosomes to be used as cytological markers in carrying out genetical analyses involving aneuploids. Telocentric "marker" lines now exist for each of the chromosomes of wheat.

II. THE USE OF ANEUPLOIDS IN GENETIC ANALYSIS

A. *The Development of Inter-varietal Chromosome Substitution Lines*

Aneuploids have been used extensively in wheat to determine the chromosomal location of genes responsible for qualitative differences (Sears, 1953). The use of monosomes (2n = 41) is particularly effective in carrying out this analysis, since qualitative gene differences are not influenced to any extent by the reduction in chromosome dosage associated with monosomy. When quantitative characters are considered, often involving a number of genes, the effect of chromosome dosage may be greater than the effect of the allelic differences being investigated. The use of monosomes and their hybrids to study quantitative inheritance has, therefore, proved to be disappointing.

To overcome the problem of chromosome dosage, another technique has been used in which single chromosomes from a donor variety can be substituted for their homologues in a second recipient variety (Sears, 1953; Law and Worland, 1973). This technique is illustrated and described in Figure 1 and involves a number of backcross generations combined with cytological selection in order to re-establish the genetic background of the recipient variety whilst maintaining the substituted chromosome intact. Because monosomics and

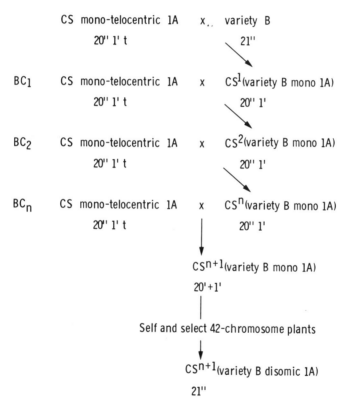

FIGURE 1. *Steps involved in producing an inter-varietal chromosome substitution line of chromosome 1A using a mono-telocentric line. Selection is practised at each backcross generation for monosomic plants. This ensures that the substituted chromosome is maintained intact and in the hemizygous condition until the final selfing generation when disomics are selected.*

telocentrics occur for each of the chromosomes of wheat, it is possible to produce the complete set of 21 inter-varietal chromosome substitution lines in which each of the chromosomes of one variety have been substituted separately for their homologues in a common recipient variety. The genetical difference between two varieties of wheat can thus be "dissected" chromosome by chromosome. Moreover, each substitution line is as true-breeding as any variety of wheat, so that is it possible to replicate each of the lines and investigate with an increased precision those characters which are of a quantitative nature.

B. *Within-chromosomal Variation*

Inter-varietal chromosome substitution lines provide the means for carrying out chromosome "assays" and investigating the relationships that exist between chromosomes. Undoubtedly, the description of such effects represent the summed effects of a number of gene differences distinguishing a substituted chromosome from its recipient homologue. One way of establishing the nature of these differences is to produce a single chromosome heterozygote in a homozygous background by hybridising a substitution line to its recipient variety (Unrau, 1958; Law, 1966). Differences between the derivatives of this hybrid consequently reflect the segregation of genes distinguishing the two homologues from each other. The study of these derivatives may permit the estimation of the number of genes and their effects to be determined.

A more precise method of analysis may, however, be used if the hybrid is first backcrossed to the recipient variety monosomic for the chromosome being studied. As Figure 2 indicates, the monosomic derivatives of this cross will then be hemizygous recombinant or non-recombinant for the critical chromosome. By selfing such derivatives, true-breeding products of a single chromosome heterozygote can be selected and used to detect differences between lines rather than single plants.

A further extension to this method is to produce an initial hybrid in which the substituted chromosome is combined with the homologous telocentric chromosome from the recipient variety. In this way, only one arm of the substituted chromosome is heterozygous, the other being hemizygous. Crossing-over is, therefore, restricted to just the one arm so that the study of the derivatives may indicate the arm location of genes and their position in relation to the centromere.

DEVELOPMENT OF SINGLE CHROMOSOME RECOMBINANT LINES

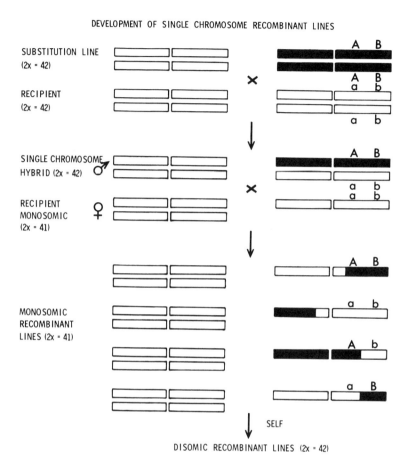

FIGURE 2. Steps involved in the development of homozygous chromosome recombinant lines from the cross of a substitution line onto its recipient variety. The donor chromosome carried by the substitution line is depicted in black and has the alleles, A and B, located on one of its arms. The recipient chromosome depicted in white carries the alleles a and b. Following the cross to the recipient monosomic, monosomic derivatives are selected which carry recombinant chromosomes in the hemizygous condition. By selfing these derivatives, disomic recombinant lines are produced.

III. THE APPLICATION OF THESE METHODS TO THE STUDY OF EAR-EMERGENCE TIME IN WHEAT

As an illustration of how these methods may be applied to the study of quantitative inheritance in wheat, use will be made of examples in which the recipient and donor varieties are Chinese Spring (CS) and Hope, respectively. Neither of these varieties are of particular agricultural merit today, but in their origins they are genetically very different and are, therefore, suitable candidates for the study of quantitative inheritance. The complete set of chromosome substitution lines of Hope into CS, produced in Missouri, U.S.A., by Dr. E. R. Sears, have also been available for some time and have been studied extensively.

Of all the quantitative characters that have been investigated, the genetical control of the days to ear-emergence time has perhaps received more attention than any other in these two varieties. The ear-emergence of Hope is earlier than CS, although both are considered as spring wheats. For the F_2 generation of the cross between them, the segregation for ear-emergence time is usually found to be of a continuous nature with a marked transgression towards lateness. Under some circumstances, however, possibly relating to different sowing dates, discontinuous variation in ear-emergence time can be observed, so that it is evident that a gene or genes having large effects relative to the effects of other genes can play a part in the control of ear-emergence time in these two varieties.

A. *Chromosome Effects*

The variations in ear-emergence time among the 21 Hope into CS substitution lines have been reported by several workers (Kuspira and Unrau, 1957; Halloran and Boydell, 1967; Law, 1968). The differences observed in one experiment in which the substitution lines along with CS and Hope were grown at $15^{\circ}C$ under continuous light in a controlled environment cabinet are given in Table II. Ten of the lines are either significantly earlier or later than CS. Of these, the most striking are those in which chromosomes 5A, 5D, and 7B of Hope have replaced their homologues in CS. Evidently, 5A and 7B of Hope carry alleles for earliness, whereas, and very strikingly, 5D of Hope carries an allele or alleles which makes CS (Hope 5D) later than either of the parental varieties. It is also of interest to note that two of the chromosomes, 5A and 5D, belong to the same homoeologous group so that the

TABLE II. Days to ear-emergence for the 21 single chromosome substitution lines of Hope into Chinese Spring (CS) and for Hope and CS grown under controlled environment conditions of 15°C and continuous light.

Substitution line	Days to ear-emergence^a
CS(Hope 1A)	70.5
CS(Hope 1B)	71.0
CS(Hope 1D)	60.5**
CS(Hope 2A)	59.8***
CS(Hope 2B)	72.0
CS(Hope 2D)	70.3
CS(Hope 3A)	72.5
CS(Hope 3B)	54.8***
CS(Hope 3D)	75.5
CS(Hope 4A)	78.0**
CS(Hope 4B)	75.3
CS(Hope 4D)	71.8
CS(Hope 5A)	54.0***
CS(Hope 5B)	61.8**
CS(Hope 5D)	92.0***
CS(Hope 6A)	71.0
CS(Hope 6B)	62.0**
CS(Hope 6D)	73.0
CS(Hope 7A)	75.5
CS(Hope 7B)	55.0***
CS(Hope 7D)	61.0**
Hope	55.9***
Chinese Spring (CS)	71.5

^{a}Significantly different from CS: *, P 0.05-0.01; **, P 0.01-0.001; ***, P < 0.001.

allelic variation observed may be functionally related even though the effects of 5A and 5D of Hope are opposed.

Similar patterns of chromosomal effects for the Hope substitution lines have been observed following a spring sowing in the field at Cambridge. The magnitude of the differences were not as great as those produced under controlled environmental conditions, but nevertheless the same chromosomes were involved with 5D again having the largest effect.

The conclusions from these studies are that many chromosomes are implicated in the control of this character and that the magnitude of their effects varies considerably. At one end of the spectrum of variation is 5D, followed by 5A and 7B, and at the other are chromosomes like 1D and 7D with smaller but nevertheless detectable effects. It is likely that the large effects of 5D are responsible for the discontinuities observed under certain conditions for the F_2 generation of crosses of CS with Hope.

B. *Between-chromosome Interaction*

Among the Hope substitution lines different from CS there is a marked tendency towards earliness whereas only CS (Hope 5D) is much later. This suggests that the control of this character is not simply additive and interactions occur between the genes on different chromosomes. Indeed, it would not be surprising if such interactions were not a common feature of wheat because of its polyploid nature.

Methods for detecting and estimating the effects of between-chromosome interactions have been described and involve the intercrossing of the substitution lines amongst themselves and with the donor and recipient varieties (Law, 1966, 1972; Aksel, 1967). These crosses enable a series of tests to be made which detect the presence of between-chromosome interaction and in some cases the particular chromosomes involved. They can also provide estimates of the parameters commonly used to describe non-allelic interactions in biometrical genetics (Mather and Jinks, 1971).

In the case of CS (Hope 5D), crosses between this line and CS and Hope, as well as the F_1 between the two varieties, provide four genotypes which can be used to detect interactions between 5D and the CS-Hope background. For example, if 5D is defined as A-a, and the genetic backgrounds of CS and Hope as B-b, then the following comparison should be zero if the effects of 5D are independent of the background.

AABb	–	AaBb	–	AAbb	+	Aabb
CS(Hope 5D) × Hope		Hope × CS		CS(Hope 5D)		CS(Hope 5D) × CS

This comparison has been made under both field and controlled environment conditions (Law, 1968, 1972) and in every case the departures from zero were significant, indicating the presence of between-chromosome interactions. Additional experiments in which CS (Hope 5A), CS(Hope 5D) and CS have been inter-crossed to give F_1 and further generations have demonstrated that a major part of the 5D between-chromosome interaction concerns the homoeologous chromosome 5A, and that this interaction is of a duplicate nature (Law, 1972; Snape et al., 1979). This type of interaction is expected where the chromosomes belong to the same homoeologous group and reinforces the suggestion already mentioned that the effects of 5A and 5D probably concern functionally identical genes.

C. Within-chromosomal Differences

Detailed analyses of true-breeding recombinant chromosome lines derived from crosses between CS and the substitution lines, CS(Hope 7B), CS(Hope 5D) and CS(Hope 5A), using the methods described in II. B, have been carried out to determine the numbers of genes responsible for the differences in ear emergence.

For CS(Hope 7B) a series of experiments were undertaken using derivatives from CS × CS(Hope 7B), as well as those from crosses involving the short and long-arm telocentric chromosomes of CS 7B. The results of these experiments are described in Figure 3, and clearly indicate from the bimodality of the distribution of the derivatives of the straight hybrid the segregation of a single genetic factor having a large effect on the character (Law, 1966). In the experiments with the telocentric chromosome derivatives, this bimodal segregation was only found for the $7B^S$ derivatives and not for those involving $7B^L$ in their development (Law and Wolfe, 1966; Law and Worland, 1973). This indicates that the short-arm is the location of this genetic factor and that it maps at 25 ± 6 units from the centromere. In subsequent experiments this genetic factor, designated $Vrn5$, has been shown to respond to low temperature or vernalisation treatments (Law, 1968; Law et al., 1976).

The analysis of the differences between the 7B homologues of Hope and CS is not however completed by the identification of $Vrn5$. The variation in ear-emergence time within each of the modes still contains a heritable component amounting to 10% of the total genetic variation. The source of this variation was shown to be the long-arm of 7B, since the derivatives of the cross involving $7B^L$ were significantly different from each other and most of this residual genetic variation could be associated with the segregation of a gene (ml) for mildew resistance located on the long-arm (Law and Wolfe, 1966). Because the derivatives of long-arm segregation gave ear-emergence values which were continuous it was not possible to determine whether the effects were due to one or more genes.

Similar investigations have been carried out on CS(Hope

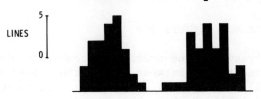
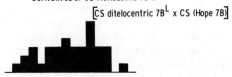
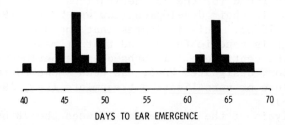

FIGURE 3. *Mean days to ear emergence for lines derived from crosses of Chinese Spring (CS), CS(Hope 7B), CS × CS(Hope 7B), CS ditelocentric 7B^L × CS(Hope 7B) and CS ditelocentric 7B^S × CS(Hope 7B) onto CS monosomic 7B.*

Segregation for a gene having a large effect on ear-emergence is shown for lines derived from the crosses of CS with CS(Hope 7B), and CS ditelocentric 7B^S with CS(Hope 7B). This segregational pattern demonstrates that a gene, Vrn5, is located on the short-arm of chromosome 7B (Law, 1966).

5D) and CS(Hope 5A). The large effects on ear-emergence associated with the substitution of these chromosomes for their homologues has been shown to be determined by single genetic factors located distally on the long-arms of each of these homoeologous chromosomes (Law et al., 1976). Again these factors respond to vernalisation and have been designated *Vrn1* for the gene on 5A and *Vrn3* for the gene on 5D. The similar locations of these genetic factors establishes almost certainly that they are identical loci, duplicated as a result of the polyploidisation of wheat, and influence the same developmental processes leading to ear-emergence.

IV. ENVIRONMENT AND GENETICAL ANALYSIS

As must be evident from the designation of these ear-emergence genes as *Vrn1*, *Vrn3* and *Vrn5*, environment can have a major influence on the magnitude of the observed genetical variation. That this should be the case for ear-emergence time is not surprising since the onset of ear initiation and subsequent stem elongation are well known to be regulated by environmental "switches" such as day-length and vernalisation. The appropriate selection of environments which maximise gene differences is, therefore, a useful addition to any genetical analysis and such selection need not just be relevant to characters such as ear-emergence time. Gene-environment interactions are frequently reported for most plant characters so that the choice of environment should be an important consideration in genetical analysis.

The influence of environment is well illustrated by a study of homozygous "recombinant" derivatives of the cross CS with CS(Hope 5D). These derivative lines were classified initially as homozygous for either *Vrn3* (the early allele) or *vrn3* (the late allele) in a controlled environment under conditions of continuous light and constant $18^{o}C$. The classified lines were then sown in the field at Cambridge in the winter and as two sowings in the early and later spring. From the winter sowing no distinction could be made between the *Vrn3* and *vrn3* lines, but in the later sowings, the difference became gradually more apparent until the lines from the late spring sowing could be separated readily into the two groups (see Figure 4).

Although this illustration exploits one particular environmental variable, temperature, controlled day-lengths, qualitative and quantitative changes in plant nutrients, changes in the amounts or types of plant infection, as well as exogenous hormonal treatments (see section V) are all environments

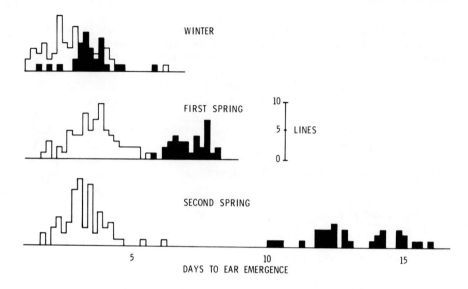

FIGURE 4. *Days to ear-emergence from commencement of emergence in the population for homozygous recombinant lines derived from the cross of Chinese Spring (CS) with CS(Hope 5D), sown at three different times, winter, first spring and second spring. The lines had previously been classified as carrying either Vrn5 (early) or vrn5 (late) alleles of a gene sensitive to low temperatures or vernalisation.*

The distribution of the lines carrying Vrn5 are shown in black, and the distribution of the lines carrying vrn5 in white.

which may usefully be considered in increasing the efficiency of genetic analysis. Indeed, such manipulations of the external environment and the interactions they produce are exactly analogous to the manipulations of genetic backgrounds which may be used to exploit genetical interactions in the identification of particular gene effects. Thus, the duplicate interactions that occur between *Vrn1* and *Vrn3* ensure that the large effects of *vrn3* on Hope 5D would not have been so apparent if allelic variation at this locus had been observed solely on a genetic background homozygous for the "early" allele *Vrn1* on Hope 5A.

V. CHARACTER ANALYSIS AND THE GENETICS OF SEMI-DWARFISM IN WHEAT

The refinements of genetical analysis described in the previous sections depend upon the precise manipulation of chromosomes between wheat varieties and the appropriate exploitation of particular environments. Yet, a further refinement is to attempt the analysis of the quantitative character itself in order to identify components which can more readily be associated with specific genes. Just as the identification of a gene affecting a particular character helps in understanding the mechanisms controlling that character, then the recognition of those mechanisms underlying a particular character expression can help in the identification of the genes responsible.

This combination of character analysis with genetic analysis has been used successfully by Thoday and his colleagues in *Drosophila melanogaster* (Thoday, 1961; Spickett, 1963) and in mouse (Spickett *et al.*, 1967; Shire, 1979). For the study of ear-emergence time in wheat obvious component characters are the time of ear initiation and the duration of stem elongation, but perhaps one of the best examples of the application of character analysis to wheat concerns the identification of the genes responsible for the successful semi-dwarf wheats of agriculture (Borlaug, 1968). These "dwarfing" genes were introduced into North America from the variety Norin 10 in the 1950s and very rapidly began to make an impact on world wheat production. However, early attempts to determine the numbers and chromosomal location of the genes involved were unsuccessful because in addition to the "dwarfing" genes final plant height is influenced by many other genes. Frequently the distribution of plant heights in F_2 and later generations were found to be continuous rather than discrete. Also, monosomic analyses (see section II A) were unsuccessful because final plant height differences were influenced greatly by chromosome dosage effects. In one such analysis at least eleven chromosomes could have been considered as carriers of the genes for semi-dwarfism (Allan and Vogel, 1963).

The first steps in character analysis were made when coleoptile length, a character correlated with final plant height, allowed the two genes conjectured from earlier height analyses (Allan *et al.*, 1968) to be identified by using a simple test cross procedure (Allan, 1970). However, coleoptile length was still not a character amenable to monosomic analysis, nor was it always definitive in identifying the genes. The final steps in the character analysis were accomplished when Norin 10 and related semi-dwarf wheats were shown to be insensitive to exogenous gibberellic acid (GA), the opposite of most other wheats where the application of GA gives rise to a marked elongation of stems and leaves. This insensitivity to GA enabled the semi-dwarfism of Norin 10 to be assigned unequivocally to two genes, *Gai/Rht1* and *Gai/Rht2*.

This analysis has, therefore, resulted in the redefinition of the final plant height character in terms of a qualitative difference, response to GA. Moreover, this difference could be established at the seedling stage so that the presence or absence of these genes could be determined and selection carried out during the early stages of any breeding programme. The character, insensitivity to GA, is also not affected by chromosome dosage so that using monosomic analysis it was possible to locate *Gai/Rht1* on chromosome 4A (Gale and Marshall, 1976) and *Gai/Rht2* on 4D (Gale et al., 1975) and also using appropriate telocentric chromosomes to show that the genes map in homoeologous positions (McVittie et al., 1978).

Undoubtedly, improvements in the physiological and biochemical understanding of plant processes will provide further opportunities for character analysis and its exploitation in genetical studies. The extent to which these advances may be used in studying the genetics of quantitative characters will depend upon the ease with which such characters can be assessed. However, even when such assessments are difficult, the ability to manipulate whole and parts of chromosomes between lines rather than just single plants of wheat should permit further refinements of the genetics of quantitative characters.

VI. PLEIOTROPIC EFFECTS OF GENES IDENTIFIED BY GENETICAL AND CHARACTER ANALYSES

The isolation of single genetic factors affecting a particular quantitative character of wheat enables the effect of that factor on other quantitative characters to be assessed. Such pleiotropic effects are of interest to wheat breeders in predicting the consequences of particular genetic substitutions as well as to physiologists, where these effects may provide an unequivocal test of causation between two or more characters. They may also lead to improved character definitions which allow particular genetic factors to be recognised more readily.

However, this approach has to face the possibility that a correlation of two or more characters with the segregation of a genetic factor, identified by the procedures described here, may be due to tightly linked genes rather than the effects of

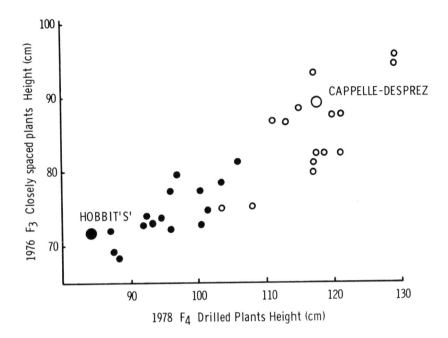

FIGURE 5. *Final plant height measurements for related F_3 and F_4 lines from the cross of Cappelle-Desprez with Hobbit sib and previously classified as carrying either Gai/Rht2 (closed circles) or gai/rht2 (open circles) in the homozygous condition. In the F_3 generation height measurements were made on spaced plants in the field (10 cm between plants) and in the F_4 generation on plants drilled at normal agricultural densities.*

a single gene. For the plant breeder such a distinction may not always be important, but for physiological research the presence or absence of linkage may be critical.

The study of the genes *Gai/Rht1* and *Gai/Rht2* provide an example of such a problem. The genes have been identified by their lack of response to GA. Clearly from Figure 5 there is a close association between GA insensitivity and reduced plant height for *Gai/Rht2*. However, this association, despite the relatively large number of lines studied, could still be due to close linkage between two genes *Gai2* and *Rht2*.

A definitive proof of a single gene would be to induce a mutation affecting both characters. However, in the absence of such evidence, other arguments may be advanced which suggest that a single gene hypothesis is more realistic. First, the GA insensitivity of plants carrying these genes probably arises from a blockage of the plant's response to endogenous GA, so that high levels of this hormone are found in the plant (Radley, 1970). This suggests that semi-dwarfism is a consequence of this blockage since an excess of GA would normally be expected to produce tall plants. Second, both *Gai/Rht1* and *Gai/Rht2* as well as a third gene, *Gai/Rht3*, identified from another source, are all GA insensitive phenotypes associated with reduced height. Such an association in three different genetical situations would seem so improbable on the basis of two different genes in each case, that pleiotropy must be considered as the most likely explanation.

The pleiotropic effects of the *Gai/Rht* genes of Norin 10 on other characters have been studied in detail. These have established that these genes improve plant yields over and above their effects on lodging resistance, by increasing the numbers of grains per ear. Associated with these increases is a not fully compensating decrease in the size of grains and a reduction in grain protein levels (see Table III and Gale *et al.*, 1979). Again the similar behaviour of each of these genes with respect to these characters can be adduced to support a pleiotropic hypothesis.

The genetic factor, *Vrn3* on 5D, identified by the aneuploid analyses described earlier (section IV) has also been studied for pleiotropic effects on other characters. It will be recalled that the "recombinant" homozygous lines, carrying either *Vrn3* or *vrn3*, were sown at three different times to

TABLE III. *The means of parents and of groups of F_3 lines classified as either Gai/Rht2 (17 lines) or gai/rht2 (15 lines) for the cross of Cappelle-Desprez (gai/rht2) with Hobbit sib (Gai/Rht2) for a range of agronomic characters.*

	Parents		F_3		Difference
	Hobbit sib	Cappelle-Desprez	Gai/Rht2	gai/rht2	(Gai/Rht2-gai/rht2)
Height (cm)	73.2	87.1	74.6	85.6	-11.0***[a]
Grain no/ear	64.0	44.5	57.9	45.4	12.5***
100 grain wt (gm)	3.9	4.6	3.9	4.7	- 0.8***
Yield/plant (gm)	20.7	18.1	19.0	17.6	1.4*
Yield/ear (gm)	2.5	2.0	2.3	2.1	0.2**
Protein %	13.5	14.6	13.5	14.7	- 1.2***

[a] *Significance levels:* *, P 0.05-0.01; **, P 0.01-0.001; ***, $P < 0.001$.

study the effect of sowing date on ear-emergence time. For the two spring sowings differences between $Vrn3$ and $vrn3$ were also found to be correlated with differences in yield, height, tiller number, grain size and grain protein content. However, for the winter sowing, no such correlations occurred. This strongly implicates a pleiotropic effect of $Vrn3$, possibly through the gene's effect on ear-emergence time, since one of the consequences of the winter sowing was to remove the difference between $Vrn3$ and $vrn3$ on ear-emergence.

VII. CONCLUSIONS

This description of the application of aneuploid and cytological marker methods has shown that it is possible, despite the complexities imposed by the polyploid nature of wheat, to achieve a detailed analysis of quantitative characters. As might be expected, these analyses have revealed a wide range of chromosomal effects which vary considerably in magnitude. Epistatic effects between chromosomes have been detected and the type of interaction described. In most cases studied (Law et al., 1978), including the one example mentioned in section III B, the between-chromosome interactions are those expected of a polyploid where a dose series of functionally related genes occur.

For those chromosomal effects which have been studied in detail it has been possible to identify genetic factors responsible for the major part of the chromosomal variation. The appropriate use of environments as well as the analysis of characters into their components have been shown to be valuable additions to the analysis and recognition of genetic factors affecting quantitative variables. For some of the genes described in detail in this paper, pleiotropy rather than linkage seems to be important. However, it should be noted that detailed studies not described here, of the effects of the gene $Vrn5$ on chromosome 7B on other characters, concluded that linkage rather than pleiotropy was responsible for some of the effects observed (Law, 1967). Undoubtedly linkage will have an important role in any studies of quantitative inheritance and the fact that it has not been highlighted in the examples quoted in this paper should not be regarded as meaningful. Rather, the presence of linkage on the one hand and pleiotropy on the other underlines the functional distinctions between two vernalisation genes, $Vrn3$ on chromosome 5D and $Vrn5$ on 7B.

It should be emphasized that the analyses described here have permitted the identification of a number of genes which have had and are having important consequences on the breeding and adaptation of wheat. Without the battery of techniques described in this paper it is doubtful whether such genes would have been recognised. Once identified, as the character analysis of the semi-dwarfing genes typifies so well, it is possible to "tag" these genes more readily, to follow their progress through breeding programmes and to predict the consequences of their substitution on plant phenotypes.

It ought also to be stated that one of the most important functions of these attempts to define quantitative variation in terms of single genetic factors should be to probe those mechanisms and processes that are of interest in plant physiological and developmental studies. The greater understanding that should result from this approach should in its turn have beneficial feedbacks to the geneticist in being able to define the genotype even more precisely.

REFERENCES

Aksel, R. (1967). *Genetics* 57, 195-211.
Allan, R.E. (1970). *Seiken Ziho.* 22, 83-90.
Allan, R.E., and Vogel, O.A. (1963). *Crop Sci.* 3, 538-540.
Allan, R.E., Vogel, O.A., and Peterson, J.C., Jr. (1968). *Crop Sci.* 8, 701-704.
Borlaug, N.E. (1968). *Proc. 3rd Int. Wheat Genet. Symp., Canberra, 1968,* 1-36.
Gale, M.D. (1979). *Proc. 5th Int. Wheat Genet. Symp., New Delhi, 1978.*
Gale, M.D., and Marshall, G.A. (1976). *Heredity* 37, 283-289.
Gale, M.D., Law, C.N., and Worland, A.J. (1975). *Heredity* 35, 417-421.
Halloran, G.M., and Boydell, C.W. (1967). *Canad. J. Genet. Cytol.* 9, 632-639.
Kuspira, J., and Unrau, J. (1957). *Canad. J. Plant Sci.* 37, 300-326.
Law, C.N. (1966). *Genetics* 53, 487-498.
Law, C.N. (1967). *Genetics* 56, 445-461.
Law, C.N. (1968). *Proc. 3rd Int. Wheat Genet. Symp., Canberra,* 331-342.
Law, C.N. (1972). *Heredity* 28, 169-179.
Law, C.N., and Wolfe, M.S. (1966). *Canad. J. Genet. Cytol.* 8, 462-470.
Law, C.N., and Worland, A.J. (1973). *Plant Breeding Institute Annual Report for 1972,* pp. 25-65.
Law, C.N., Worland, A.J., and Giorgi, B. (1976). *Heredity* 36, 49-58.

Law, C.N., Sutka, J., and Worland, A.J. (1978). *Heredity 41*, 185-191.
Mather, K., and Jinks, J.L. (1971). "Biometrical Genetics." Chapman and Hall Ltd., London.
McVittie, J.A., Gale, M.D., Marshall, G.A., and Westcott, B. (1978). *Heredity 40*, 67-70.
Radley, M. (1970). *Planta 92*, 292-300.
Sears, E.R. (1953). *Amer. Nat. 87*, 245-252.
Sears, E.R. (1954). *Missouri Agric. Exp. Sta. Res. Bull. No 572.* pp. 59.
Sears, E.R. (1966). *In* "Chromosome Manipulations and Plant Genetics" (R. Riley and K.R. Lewis, eds.), pp. 29-45. Oliver and Boyd, Edinburgh.
Shire, J.G.M. (1979). *In* "Quantitative Genetic Variation" (J.N. Thompson, jr. and J.M. Thoday, eds.), pp. 263-273. Academic Press, New York.
Snape, J.W., Law, C.N., Young, C.F., and Worland, A.J. (1979). *Heredity 42*, in press.
Spickett, S.G. (1963). *Nature 199*, 870-873.
Spickett, S.G., and Thoday, J.M. (1966). *Genet. Res. 7*, 96-121.
Thoday, J.M. (1961). *Nature 191*, 368-370.
Unrau, J. (1958). *Canad. J. Plant Sci. 38*, 415-418.

SYNTHESIS: POLYGENIC VARIATION IN PERSPECTIVE

James N. Thompson, jr.

Department of Zoology
University of Oklahoma
Norman, Oklahoma

J. M. Thoday

Department of Genetics
University of Cambridge
Cambridge, England

The chapters in this book focus both upon the techniques used to study variation and, to a lesser extent, upon the nature of genetic variation itself. Some techniques, such as the location and manipulation of polygenic loci, are clearly applicable to only a limited array of species for which especially precise breeding programs are possible. Other techniques, such as the comparison of isofemale strains, biometrical analysis, the study of selection responses, and recombinant inbred lines can be used in a wide variety of organisms.

Some approaches give only a general idea of the nature of the underlying genetic variation, though many purposes do not require more. Indeed, to some extent the perfection of analytical methods has preceeded any clear understanding of the nature of the genes involved and sometimes it is even forgotten that the deficiency exists. Thus, it might be appropriate to use this final chapter to consider briefly the unknowns at the foundation of quantitative genetic variation: how many genes are involved, what sort of genes are they, and what do they do?

Although popular conception sometimes assumes large numbers of genes contributing to continuous phenotypic distributions, this is not a necessary consequence of either biometrical analysis or selection theory (see, for example, chapter by

Mather). As Thoday and Thompson (1976) point out, some of the confusion surrounding gene number derives from the fact that different people mean different things by the question "How many genes are involved?"

The most general version of this question is "How many genes are involved in influencing variable x in species y?" This includes all genes in the gene pool, both homozygous and heterozygous, that contribute directly or indirectly to the development or expression of the character. The second version, "At how many loci in species y, or population y_1, is there allelic *segregation* affecting variable x?", draws attention to the smaller number of segregating loci. A third version focuses upon the genes responsible for the *major* part of the variation, while a final version, more precise than the others, asks "At how many loci is there segregation *directly* affecting the development of the character?" It is these last two versions that point the direction in which a deeper understanding of the relationship between individual development and population genetic structure can be sought.

Although very little is known about the types of gene effects subsumed in the category "polygenes", several guides to future studies can be drawn from the literature. Perhaps the best idea is to consider the types of gene effects that *might* theoretically be involved. We are certain that it is not an exhaustive list, but it will at least provide a point of departure for discussions of gene action.

First, there is little evidence that the number of polygenes affecting any particular trait is exceptionally large. Indeed, in those instances in which gene number has been studied in detail, a very limited number of loci account for the great majority of the genetic variation (*e.g.*, review by Thompson, 1975a, 1976). Thus, polygenes are not likely to be qualitatively different from other genes. This leads naturally to the hypothesis that polygenic variation can be due to the segregation of isoalleles of loci also known through mutant alleles having a larger phenotypic effect. Though isoalleles are not always implicated (Waddington *et al.*, 1957), there are several useful examples of isoallelic contributions to selection response. Milkman (1970) found an allele of *detached* in a *Drosophila* line selected for reduced crossveins, and Scharloo (1970) found two closely-linked loci having opposite effects upon vein length in a ci^{D-G} selection line. One was an allele of *crossveinless-2* and the other was an allele of *plexus*. The involvement of isoalleles is not always direct and predictable, however, as Thompson (1975b) has shown.

The involvement of control mutations has been proposed by several authors to explain quantitative variation. Pandey (1972), for example, has suggested that there is a functional distinction between polygenes and "major genes" and that polygenes act by directly or indirectly controlling major genes. Others, including Rendel (1968), have suggested that polygenes support the function of major genes either by doing the same thing as the major gene or by making the cellular environment favorable for the action of the major gene or its products. An insight into some of the ways such variation might act is provided by variants for enzyme activity mapping near a gene (*e.g.*, the XDH enzyme locus, *rosy*, Chovnick et al., 1976, 1977; the *Adh* locus, Thompson et al., 1977) and at clearly unlinked loci (*e.g.*, McDonald and Ayala, 1978; Barnes and Birley, 1978). However, the identification of control or regulatory genes is difficult in eucaryotes, and critical evidence is limited.

In a similar way, random inactivation of loci in cells or patches of tissue, as in position effect variegation (Spofford, 1976), leads to phenotypically variable individuals, particularly if a suppressor or enhancer of variegation is also segregating in the population. Both this and the prediction of control gene effects lead one to expect locus-specific polygene action. Though these are undoubtedly present in some situations, there are two other classes of gene effects that are not locus-specific.

The first of these is variation in the protein synthetic machinery of the cell, particularly variations in rRNA and tRNA. Variation in ribosomal RNA has recently been implicated in *Drosophila* bristle selection lines (Frankham et al., 1978), and Hillman (*e.g.*, 1977) showed that the residual (polygenic modifier) genotype that was responsible for increased penetrance and expressivity of the sex-linked gene *Abnormal abdomen* in *Drosophila melanogaster* was also responsible for an increased aminoacylating enzyme activity of tRNA. One might predict that such contributors to variation might produce a wide range of phenotypic correlations.

At the other end of the scale, the lack of correlations among systems, such as bristle traits (*e.g.*, Davies and Workman, 1971), suggests that there is some degree of independence among the developmental processes leading to phenotypic variation in superficially similar traits. This should not surprise us. Afterall, the structures are being formed in quite different parts of the body and are, therefore, in possibly quite different cellular environments. Indeed, the minor variation in developmental effects provides a powerful tool for studying the subtle gene interactions that yield patterns of structures, such as bristle or vein arrangements (see, for example, the extensive work by F.W. Robertson and collaborators, *e.g.* Masry and Robertson, 1978; chapter by Thompson).

The dissection of developmental effects contributing to

variation also leads us into a more sophisticated view of the
nature of quantitative traits. Undoubtedly the most elegant
study is that done by Spickett (1963) on variation in sternopleural bristle number in *Drosophila melanogaster* (Figure 1).
In collaboration with Thoday, Spickett located three polygenes
that accounted for the majority of the phenotypic difference
between a high and a low bristle number selection line (see
chapter by Thoday). A fourth gene affecting fly size was also
involved. These four genes were isolated and their developmental effects were identified as summarized in Figure 2. The
picture they provided is one in which a character can be seen
as a complex interaction of general processes such as cell
division and growth rates and of more character-specifc processes such as bristle precursor distribution. Variation in
any of these would be classified as due to the segregation of
"bristle number polygenes", though only a proportion would
have direct influence upon the developmental processes of
bristles alone.

This discussion of possible types of polygenes leads us to
a second major question. What function does polygenic variation play, particularly in traits such as wing vein length or
scutellar bristle number, which are essentially invariant in
natural populations? Such a question can perhaps take a
lesson from comparing two prevalent hypotheses to explain
electrophoretic polymorphism: selection in heterogeneous environments, and the supposed inherent advantage of hybrid
enzyme molecules. Although there is some evidence that heterogeneous environments favor genetic variability, there is

FIGURE 1. *Sternopleural bristles between the first and
second legs of an adult Drosophila melanogaster.*

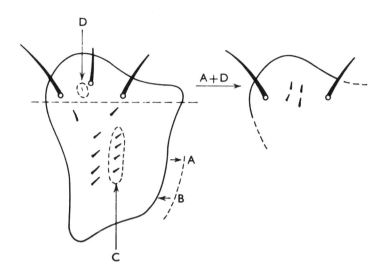

FIGURE 2. *The use of genes in the analysis of a complex character. The effects of four genes isolated from a high sternopleural hair number line of Drosophila. The left-hand figure shows the pattern of hairs on the sternopleurite of wild-type Oregon inbred and, enclosed in dotted lines, the effects of individual genes. The right-hand figure shows the effect of genes A and D together. A increases cell number and hair number in all regions. B decreases cell size, compensating for the fly size effect of A without influencing hair number. C has a local effect on hair number tending to give a double row of microchaetes. D has a very local effect adding one hair near the middle macrochaete. With A it replaces this macrochaete by several microchaetes. It works by delaying differentiation of the chaetal initial of this middle macrochaete. Increase of cell number in this region then leads to differentiation of several initials which form microchaetes. (From Thoday, 1967).*

little or no evidence to support the hypothesis that hybrid multimeric enzymes are inherently superior. In fact, neither hypothesis is completely satisfactory.

As Johnson (1976) points out, however, the problem is that these hypotheses are single-locus conceptualizations that consider gene products as acting independently of eachother. But this is definitely inconsistent with the evidence from molecular biology, which shows that the phenotype is a coordinated product of many reactions. Much of developmental homeostasis can be hypothesized to be due to a series of feedback mechanisms, rate dependent reactions, intercellular interactions such as growth rate compensation, and other means of controlling and coordinating the interacting contributions of development.

Genetic variation might, therefore, contribute to developmental homeostasis by expanding the range of environmental conditions within which efficient developmental coordination can occur. At individual loci, this view does not differ from the single-locus explanation of electrophoretic polymorphism. But consideration of traits such as weight, I.Q., bristle number, and wing vein length, makes it much easier for us to see the limitations of a single-locus perspective. The same coordination, when expanded to encompass an entire developmental process, increases the power and gives us a hint of the adaptive significance of polygenic heterogeneity.

The study of quantitative genetic variation remains a dynamic and challenging field. Many technical approaches to variation have been applied to a wide variety of questions. We hope that this survey of methodologies will help put some of the current uses and problems into perspective and contribute to further synthesis among the fields that touch, or are touched by, quantitative genetics.

REFERENCES

Barnes, B.W., and Birley, A.J. (1978). *Heredity 40*, 51-57.
Chovnick, A., Gelbart, W., McCarron, M., Osmond, B., Candido, E.P.M., and Baillie, D.L. (1976). *Genetics 84*, 233-255.
Chovnick, A., Gelbart, W., and McCarron, M. (1977). *Cell 11*, 1-10.
Davies, R.W., and Workman, P.L. (1971). *Genetics 69*, 353-361.
Frankham, R., Briscoe, D.A., and Nurthen, R.K. (1978). *Nature 272*, 80-81.
Hillman, R. (1977). *Amer. Zool. 17*, 521-533.
Johnson, G.B. (1976). In "Molecular Evolution" (F.J. Ayala, ed.), pp. 46-59. Sinauer, Sunderland, Mass.
McDonald, J.F., and Ayala, F.J. (1978). *Genetics 89*, 371-388.
Masry, A.M., and Robertson, F.W. (1978). *Egypt. J. Genet. Cytol. 7*, 137-165.
Milkman, R.D. (1970). *Adv. Genet. 15*, 55-114.
Pandey, K.K. (1972). *Theoret. Appl. Genet. 42*, 250-261.
Rendel, J.M. (1968). In "Evolution and Environment" (E.T. Drake, ed.). Yale Univ. Press, Yale.

Scharloo, W. (1970). *Genetics 65*, 681-691.
Spickett, S.G. (1963). *Nature 199*, 870-873.
Spofford, J.B. (1976). *In* "The Genetics and Biology of Drosophila, Volume 1c" (M. Ashburner and E. Novitski, eds.), pp. 955-1018. Academic Press, London.
Thoday, J.M. (1967). *In* "Endocrine Genetics" (S.G. Spickett and J.G.M. Shire, eds.), pp. 297-309. Mem. Soc. Endocrin. no. 15. Cambridge Univ. Press, Cambridge.
Thoday, J.M., and Thompson, J.N., jr. (1976). *Genetica 46*, 335-344.
Thompson, J.N., jr. (1975a). *Nature 258*, 665-668.
Thompson, J.N., jr. (1975b). *Heredity 35*, 401-406.
Thompson, J.N., jr. (1976). *Nature 261*, 525-526.
Thompson, J.N., jr., Ashburner, M., and Woodruff, R.C. (1977). *Nature 270*, 363.
Waddington, C.H., Graber, H., and Woolf, B. (1957). *J. Genet. 55*, 246-250.

Index

A

Accelerated response, 225
Additive variation, 11, 83
Adrenal tissue, 265, 268
Aggression, 207
Alcohol, 61, 68–71, 74, 208–209
Alcohol dehydrogenase, 24, 26, 69, 75
Allele frequencies, 132
Allelism test, 166
Analysis of variance, 184
Aneuploid, 275–293
Animal breeding, 28
Ascospore size, 41
Aspergillus nidulans, 37, 41–42, 46–47, 49–54

B

Balance, 86
Behavior, 74–75, 208–211, 266, 269
Biological diversity, 71
Biometrical genetics, 10–15, 81–109
Bristles
  abdominal, 17, 20, 25, 185–186, 192–193
  coxal, 22
  scutellar, 61
  sternopleural, 17, 20, 185, 192–193, 221–229, 298–299

C

Canalization, 139–156
Character, redefinition, 263

Chromosome
  assay, 10, 81
  effects, 148, 280–282
  identity test, 167–171
  location, 69, 73
  mapping, 169
  recombinant, 16, 279
  substitution, 236, 276–279
Chromosome B, 24
Chromosome X, 25
Chromosome Y, 25
^{60}Co-γ irradiation, 72
Collybia velutipes, 40, 42–43, 49
Complementation, 26
Computer simulation, 121–137, 235–242
  advantages, 135
Correlated responses, 47, 116
Correlation, 247
Corticosteroid, 207, 264–265, 267, 271
Covariances, 12, 14
Crossvein, 140, 157–176

D

Dacus tryoni, 68
Desiccation, 61, 65
Development, 11, 220, 243–261
Developmental noise, 163, 165–166
Diallel crosses, 98, 100
Dominance, 11, 15, 30, 83, 102
Dosage compensation, 153–154
Drosophila, 9, 11, 16, 20–22, 24–25, 27–30, 61–79, 139–156, 157–176, 177–196, 220–229, 243–261, 296–299

E

Effective factors, 17–20, 102, 231–233
  location, 50–51
  number, 48–53
Electrophoretic markers, 170–171
Environment, 83, 131, 220–221, 241, 285–286
Epistasis, 14–15, 40, 130
Experimental design, 96–101
Exploratory activity, 209–211, 213

F

Fungi, 35–59
  experimental properties, 35

G

Gene
  effects, 16, 20–23, 42, 129–130
  evolution, 20, 22, 26
  location, 16, 151
  number, 13, 17–18, 15–20, 239, 256, 295–298
Genetic
  assimilation, 162, 173–174
  gain, 133
Gradient, 243, 259

H

Heritability, 27–28, 106, 115, 131, 133, 150
Heterochromatin, 24–25
Heterosis, 29, 104
Histocompatibility, 197–218, 269–270
  loci, number of, 199, 201

I

Identity test, 166
Illinois experiment station, 18, 27, 29
Immunogenetics, 206–207
Inbreeding, 11, 16, 18, 28–30, 112, 201, 267
  coefficient, 133
Interactions, 83–84, 94, 102–103, 282–283
Irradiation, 26, 185, 189–190, 192–193
Isoalleles, 25, 174, 220, 255
Isofemale
  lines, 161, 166
  strains, 61–62, 161, 166
Isogenic, 199
Isozyme, 24

L

Law of ancestral heredity, 6
Learning, 174, 207
  active avoidance, 211–212
Linkage, 9, 15, 19, 51, 129, 205
  disequilibrium, 30, 102, 113, 129
  repulsion, 113
Loci, polymorphic, 172
Longevity, 62, 72, 74

M

Maize (*Zea mays*), 18, 19, 27, 116, 131, 134
Mapping, 166, 171, 219–233, 235–242
Maternal effects, 15
Means, 87–90, 132
Mice, 18, 19
Mimicry, 31
Model fitting, 92–93
Mosaic, 152, 257–259
Mouse news letter, 198
Multiple mating design, 99
Mutation, 162, 177–196

N

Neurospora crassa, 37, 41, 47
Nicotiana rustica, 19
North Carolina designs, 98, 181

P

Parasexual cycle, 50
Partial diallel cross, 181
Pattern, 245, 257
Penicillin, 36, 42, 50, 54–55
Penicillium chrysogenum, 49
Phaseolus vulgaris, 6, 9, 29
Phenodeviants, 161, 166, 173
Phenotypic load, 118
Phototaxis, 74–75
Physiology, 263–273
Pleiotropy, 9, 23, 130, 205, 245, 288–291
Plexus, 248–254
Polygene, 6–7, 16–17
  location, 170, 219–233, 235–242, 247
  magnitude, 181, 239, 241
  number, 3, 123, 172, 208, 239, 256–257, 264
Population size, 125
Prepattern, 152–153, 251
Probit analysis, 142–143, 151

Progeny testing, 11, 15–16, 222, 224, 226, 228, 237–238, 240–241

R

Radial growth rate, 40, 46, 53–54
Radiosensitivity, 72
Recombinant inbred lines, 105–106, 197–218, 270
Recombination, 19, 112, 162
Rodents, 263–274
Rye (*Secale cereale*), 24, 134

S

Scale, 21, 84–86
Scaptomyza, 74
Schizophyllum commune, 37, 40, 42–43, 46–47
Scute, 141, 144, 146–154
Segregation, 8–9, 221
Selection, 18–19, 27–31, 44–47, 81, 106, 111–119, 139–156, 166, 246–253
  differential, 114–115
  directional, 114–116
  disruptive, 118–119
  intensity, 115, 118, 125
  limits, 47
  stabilizing, 117–118
Selfing, 92
Sex-linkage, 15

Sorghum, 24
Strain distribution pattern (SDP), 198
Subthreshold, 251–252

T

Tabby, 141, 145–146
Temperature, 61
  sensitivity, 52–53
Tetrads, 40–41
Thresholds, 142, 158–161, 219, 251–252
Trait, choice of, 157, 244–246
Triple test cross, 86, 99–100, 104

U

Ustilago hordei, 41, 43

V

Variances, 90–92, 133
Variation, hidden, 112
Veinlet, 246–248
Viability, 179–182, 187–190
Vibrissae, 140–141, 145

W

Wheat (*Triticum aestivum*), 275–293
Wing veins, 244–261

79486　　　　　　　　　　　　　QH
　　　　　　　　　　　　　　　452.7
　　　　　　　　　　　　　　　Q36

QUANTITATIVE GENETIC VARIATION

DATE DUE	
NOV 0 2 1998	
GAYLORD	PRINTED IN U.S.A.

Fernald Library
Colby-Sawyer College
New London, New Hampshire